LONDON MATHEMATICAL SOCIETY LECTURE NOTE SERIES

Editor: PROFESSOR G. C. SHEPHARD, University of East Anglia

This series publishes the records of lectures and seminars on advanced topics in mathematics held at universities throughout the world. For the most part, these are at postgraduate level either presenting new material or describing older material in a new way. Exceptionally, topics at the undergraduate level may be published if the treatment is sufficiently original.

Prospective authors should contact the editor in the first instance.

The following titles are available

1. General cohomology theory and K-theory, PETER HILTON.
4. Algebraic topology: A student's guide, J. F. ADAMS.
5. Commutative algebra, J. T. KNIGHT.
7. Introduction to combinatory logic, J. R. HINDLEY, B. LERCHER and J. P. SELDIN.
8. Integration and harmonic analysis on compact groups, R. E. EDWARDS.
9. Elliptic functions and elliptic curves, PATRICK DU VAL.
10. Numerical ranges II, F. F. BONSALL and J. DUNCAN.
11. New developments in topology, G. SEGAL (ed.).
12. Symposium on complex analysis Canterbury, 1973, J. CLUNIE and W. K. HAYMAN (eds.).
13. Combinatorics, Proceedings of the British combinatorial conference 1973, T. P. McDONOUGH and V. C. MAVRON (eds.).
14. Analytic theory of abelian varieties, H. P. F. SWINNERTON-DYER.
15. An introduction to topological groups, P. J. HIGGINS.
16. Topics in finite groups, TERENCE M. GAGEN.
17. Differentiable germs and catastrophes, THEODOR BRÖCKER and L. LANDER.
18. A geometric approach to homology theory, S. BUONCRISTIANO, C. P. ROURKE and B. J. SANDERSON.
19. Graph theory, coding theory and block designs, P. J. CAMERON and J. H. VAN LINT.
20. Sheaf theory, B. R. TENNISON.
21. Automatic continuity of linear operators, ALLAN M. SINCLAIR.
22. Presentation of groups, D. L. JOHNSON.
23. Parallelisms of complete designs, PETER J. CAMERON.
24. The topology of Stiefel manifolds, I. M. JAMES.
25. Lie groups and compact groups, J. F. PRICE.
26. Transformation groups: Proceedings of the conference in the University of Newcastle upon Tyne, August 1976, CZES KOSNIOWSKI.
27. Skew field constructions, P. M. COHN.
28. Brownian motion, Hardy spaces and bounded mean oscillation, K. E. PETERSEN.
29. Pontryagin duality and the structure of locally compact abelian groups, SIDNEY A. MORRIS.

London Mathematical Society Lecture Note Series. 30

Interaction Models

COURSE GIVEN AT ROYAL HOLLOWAY COLLEGE,
UNIVERSITY OF LONDON,
OCTOBER-DECEMBER 1976

NORMAN BIGGS

Reader in Pure Mathematics
Royal Holloway College
University of London

CAMBRIDGE UNIVERSITY PRESS
CAMBRIDGE
LONDON NEW YORK MELBOURNE

CAMBRIDGE UNIVERSITY PRESS
Cambridge, New York, Melbourne, Madrid, Cape Town, Singapore, São Paulo

Cambridge University Press
The Edinburgh Building, Cambridge CB2 8RU, UK

Published in the United States of America by Cambridge University Press, New York

www.cambridge.org
Information on this title: www.cambridge.org/9780521217705

First published 1977
Re-issued in this digitally printed version 2008

A catalogue record for this publication is available from the British Library

ISBN 978-0-521-21770-5 paperback

Contents

1 · Preview

1.1 Apologia

The lectures on which this book is based were intended for a 'mixed audience'. According to the context, that phrase might have certain social connotations, but here it implies a more fundamental distinction: some of the audience were basically physicists, and others were basically mathematicians. This distinction, between those who think in terms of real objects and those who deal in abstract ideas, is an unfortunate fact of scientific life today.

The desire to be intelligible to two classes of student has been my main preoccupation in preparing the lectures and writing the book. Consequently, any reader will probably find some material which (to him) is tiresome and elementary; such material is included for the benefit of other readers, in the cause of scientific harmony. I have tried to prescribe a proper dose of generality - not too much to discourage those who have a particular application in mind, nor too little for those who wish to see the underlying structure.

The book has five chapters, each subdivided into sections. The first chapter is intended as a broad introduction to the subject, and it is written in a more informal manner than the rest. There are two short appendices at the end of the book, and these are referred to in Chapters 2, 3 and 4. Apart from this, there are no references in the main text; notes and references for each chapter are given at the end of the chapter.

1.2 States on a graph

Our main object of study is a finite system of particles, some of which interact in pairs. We shall not be concerned with the nature of the particles, their relative positions, or the kind of interactions; such things are important when one studies a particular physical structure, but they do not lie at the heart of the matter. In order to describe the pattern of interactions among the particles of the system we shall require a little of the terminology of graph theory.

A **graph** consists of a set of **vertices** and a set of **edges**, together with an **incidence relation**: each edge is **incident** with either one or two vertices. Graphs are often represented by diagrams in which the vertices are points and the edges are line segments joining the relevant points (see Fig. 1). An edge incident with just one vertex is called a **loop**, and is represented as such. Sets of edges incident with the same pair of vertices are called **multiple edges**. A graph with no loops or multiple edges is said to be **simple** (Fig. 1). We shall adopt the convention that the term 'graph' will always mean 'simple graph', unless the context indicates otherwise.

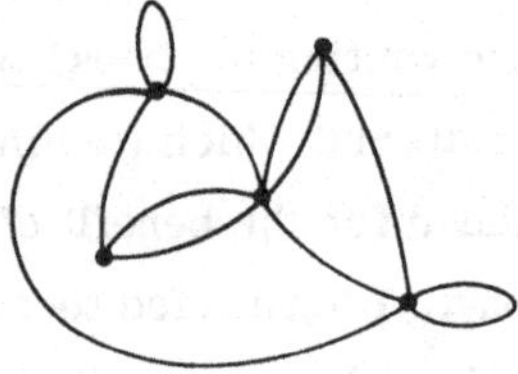

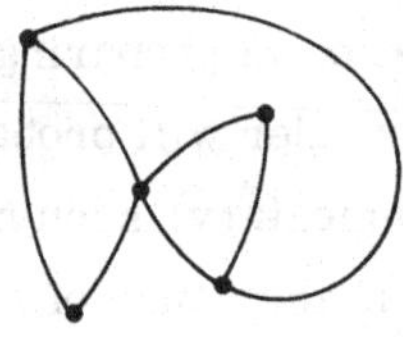

Fig. 1. (a) A graph (b) A simple graph

In the ensuing theory it will be usual to think of the vertices of a graph as a set of particles, where the edges signify the existence of interactions between some pairs of particles.

We shall reserve the notations V and E for the vertex-set and edge-set of a graph; occasionally, when the graph G needs to be named explicitly, VG and EG will be used. A graph is said to be **connected** if any pair v, w of its vertices may be linked by a **path**, that is, a sequence

$$v = v_0, e_1, v_1, e_2, \dots, e_r, v_r = w$$

of alternate vertices and edges, where e_i is an edge incident with v_{i-1} and v_i $(1 \leq i \leq r)$. If a graph is not connected, it falls into a number of **components**, each of which is connected. In terms of a system of interacting particles, we may separate the system into subsystems, each of which is independent of the rest and cannot be further subdivided.

Let A be a finite set, and G a graph. A function ω from the vertex-set V of G to the set A (written $\omega : V \to A$) assigns to each vertex an element of A; that is, each particle is given some 'attribute' or 'configuration' or 'colour'. A function ω thus defines a **state** of the system of particles represented by the graph, and the set of all such states will be denoted by $\Omega(G, A)$, or just Ω.

For theoretical reasons it is convenient to endow the set A with some algebraic structure. The most apt structure is that of a 'ring'; that is, we postulate the existence of two operations, + and $\cdot$, satisfying the usual rules of arithmetic, except that division (the inverse of the $\cdot$ operation) is not allowed. All the rings that we consider will contain elements 0 and 1, with the usual properties of those symbols, and both operations will be commutative. For each positive integer m, the set of residues modulo m forms a ring, with the usual operations of modular arithmetic. We shall denote this special ring by A_m. Thus any non-empty finite set is in one-to-one correspondence with at least one ring. It follows that a ring structure may be imposed on any finite set of attributes, without loss of generality, and with some gain. For many purposes (but not all) the rings of residues are a sufficient level of generality.

When A is a ring the states ω in $\Omega(G, A)$ may themselves be combined by operations derived from the structure of A. In fact, if ω_1 and ω_2 are such states, then we may define states $\omega_1 + \omega_2$ and $\omega_1 . \omega_2$ by the rules

$$(\omega_1 + \omega_2)(v) = \omega_1(v) + \omega_2(v) ,$$

$$(\omega_1 . \omega_2)(v) = \omega_1(v) . \omega_2(v) .$$

The set Ω thus becomes a ring itself. In particular, there is a 0-state 0 defined by $0(v) = 0$ for all v in V, where the second 0 is the zero element of A, and similarly a 1-state defined by $1(v) = 1$.

In a manner analogous to that used to define Ω, we may introduce the ring $\Phi = \Phi(G, A)$ of functions $\phi : E \to A$. We think of a function ϕ in Φ as an assignment of a 'flow' $\phi(e)$ to each edge e of G. In order to describe the 'direction' of such a flow, we must choose an **orientation** of the graph G; that is, for each edge e of G one of the two incident vertices is chosen to be the 'positive end' of e, and the other is chosen to be the 'negative end'. We make the convention that the single vertex incident with a loop is its positive end. An orientation is usually represented on a diagram by placing an arrow on each edge, pointing towards its positive end (Fig. 2). Although the introduction of

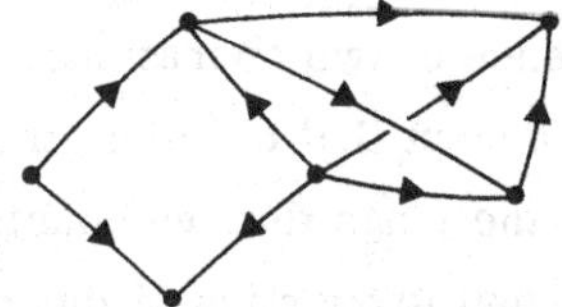

Fig. 2. A graph with an orientation

an orientation is necessary in order to yield satisfactory definitions, the actual orientation chosen is immaterial.

Given a finite graph G with an orientation, we define a matrix D with $|V|$ rows, labelled by the vertices of G, and $|E|$ columns, labelled by the edges of G, as follows:

$$D_{ve} = \begin{cases} 1 & \text{if } v \text{ is the positive end of } e; \\ -1 & \text{if } v \text{ is the negative end of } e; \\ 0 & \text{if } v \text{ is not incident with } e. \end{cases}$$

D is called the **incidence matrix** of G, with respect to the chosen orientation. The entries of D are taken to be elements of the ring A; every ring has elements 0, 1, and -1, although in the case of the residues modulo 2 (for example) 1 and -1 are the same.

We can now define two important operators. They are:

the boundary operator $\partial : \Phi(G, A) \to \Omega(G, A)$

and the coboundary operator $\delta : \Omega(G, A) \to \Phi(G, A)$.

The operator ∂ assigns to each flow ϕ on G a state $\partial\phi$, and δ assigns to each state ω a flow $\delta\omega$, defined by

$$(\partial\phi)(v) = \sum_{e \in E} D_{ve} \cdot \phi(e); \qquad (\delta\omega)(e) = \sum_{v \in V} D_{ve} \cdot \omega(v).$$

Intuitively, the value of $(\partial\phi)(v)$ represents the net accumulation of flow at the vertex v. As for $\delta\omega$, if e is not a loop then there are just two non-zero terms D_{ve} : a term D_{xe} corresponding to the positive end x of e ($D_{xe} = 1$), and a term $D_{ye} = -1$ corresponding to the negative end y of e. Consequently

$$(\delta\omega)(e) = \omega(x) - \omega(y),$$

and $(\delta\omega)(e)$ represents the difference in the values of ω at the ends of e.

We shall introduce other notions from graph theory as they are needed.

1.3 Interaction models

In this section we shall set up a mathematical structure, called an interaction model, which is the main topic in the remainder of this book. The physical background motivating the formulation of the model will be surveyed in the next section.

For the time being, all graphs considered will be simple and finite. We begin with the set $\Omega = \Omega(G, A)$ of states on a graph G with values in the ring A. To each state $\omega : V \to A$ we assign a complex number $I(\omega)$ called the weight of ω; in other words I is a function

$$I : \Omega(G, A) \to C,$$

where C denotes the complex numbers. In many instances the values of I will be real and non-negative. When that is the case we may put

$$Z = \sum_{\omega \in \Omega} I(\omega)$$

and interpret the quantity $I(\omega)/Z$ as being the probability that the particles of the system have the attributes specified by ω.

We shall be concerned with weights which are derived from the local structure of the graph G in the following way. Any state ω has a coboundary $\delta\omega$; as explained in the previous section, the value $\delta\omega(e)$ represents the difference in the values of ω at the ends of the edge e. We introduce an 'interaction function' $i : A \to C$, whose value $i(a)$ is taken to represent the strength of the interaction between two particles when the difference (in the ring A) of their 'attributes' is a. Because the interaction should not depend on the order in which the particles are considered, we shall always assume that i satisfies the symmetry condition $i(a) = i(-a)$. When the state of the system is specified by ω, the interaction on the edge e of G is given by $i[\delta\omega(e)]$; because of the symmetry condition, this does not depend on the orientation used in defining δ. We shall take the weight of ω to be the product of these terms over all edges of G. (The reason for the occurrence of a product, rather than a sum, will appear shortly.) We are now in a position to recast the foregoing discussion into a set of basic definitions.

An **interaction model** $\mathfrak{M}$ consists of a ring A and an **interaction function** $i : A \to C$, with the property $i(a) = i(-a)$. We write $\mathfrak{M} = (A, i)$, signifying that $\mathfrak{M}$ is an ordered pair.

If we are given an interaction model $\mathfrak{M}$ and a graph G, we shall speak of the interaction model $\mathfrak{M}$ on G, and associate with $\mathfrak{M}$ and G the **weight function**

$$I(\omega) = \prod_{e \in E} i[\delta\omega(e)] , \tag{1.3.1}$$

and the **partition function**

$$Z(\mathfrak{M}, G) = \sum_{\omega\in\Omega} I(\omega) = \sum_{\omega\in\Omega} \prod_{e\in E} i[\delta\omega(e)]. \tag{1.3.2}$$

For example, let us consider the very simple model $\mathfrak{C}$ whose interaction function is defined by

$$i(a) = \begin{cases} 1 & \text{if } a \neq 0, \\ 0 & \text{if } a = 0. \end{cases}$$

For this function, $i[\delta\omega(e)] = 0$ when ω assigns the same element of A to the two ends of e, and it is equal to 1 otherwise. Consequently $I(\omega)$ is zero unless ω assigns different 'colours' to the two ends of each edge in G, when it is equal to 1. If we use the above notation, then $Z(\mathcal{C}, G)$ is the number of 'proper colourings' of G with $|A|$ colours, where the adjective 'proper' signifies that adjacent vertices have different colours. This function has been much studied in graph theory, under the name of the 'chromatic polynomial'.

In physics, the models studied usually have the property that the interaction i(a) is real and positive, for all a in A. We shall say that an interaction model is **positive** if its interaction function has this property. Such interaction functions may be written as

$$i(a) = \exp j(a) \qquad (a \in A),$$

for some real-valued function j defined on A. The corresponding weight function I is then of the form

$$I(\omega) = \exp J(\omega) \qquad (\omega \in \Omega),$$

where

$$J(\omega) = \sum_{e \in E} j[\delta\omega(e)] .$$

These remarks (together with the physical principles described in the next section) explain how the sum of interactions becomes a product in our general formulation.

Interactions of the kind described in the preceding paragraph have an interesting additional property, related to the probability interpretation of the weight function. Suppose that I is any real, positive weight function on Ω, and extend I to the subsets of Ω by putting

$$I(\Lambda) = \sum_{\omega \in \Lambda} I(\omega) \qquad (\emptyset \neq \Lambda \subseteq \Omega),$$

$$I(\emptyset) = 0.$$

Then I/Z is a probability measure on the subsets of Ω, and we may define conditional probabilities in the usual way. That is, the conditional probability that a state belongs to a subset X of Ω, given that it belongs to Y, is $I(X \cap Y)/I(Y)$.

Let v be a given vertex of a simple graph G. We shall say that a real positive weight function I has the **Markov property** if the following holds: the conditional probability that a state is ω, given that it agrees with ω on all vertices of G except v, depends only on the values of ω at v and the vertices adjacent to v in G. Roughly speaking, the probability that the system has a particular configuration at v depends only on its values at the neighbours of v.

It is fairly easy to see that any weight function I, which arises from a positive interaction model $\mathcal{P} = (A, i)$, has the Markov property. In view of the fact that interactions occur only between neighbouring vertices, this is perhaps not too surprising, but a proof seems called for. First, it is clear from (1.3.1) that I is real and positive whenever i is. Now the conditional probability occurring in the definition of the Markov property is just

$$I(\omega)/\Sigma I(\theta),$$

where the sum is over all those states θ which agree with ω except at v. These are the states ω_a $(a \in A)$ defined by

$$\omega_a(v) = \omega(v) + a, \quad \omega_a(w) = \omega(w) \qquad (w \neq v).$$

In particular, $\omega = \omega_0$. Now

$$\begin{aligned} I(\omega_a) &= \prod_{e \in E} i[\delta\omega_a(e)] \\ &= (\text{product over edges } e \text{ incident with } v) \\ &\quad \times (\text{product over } e \text{ not incident with } v). \end{aligned}$$

Since ω_a is independent of a except at v, $\delta\omega_a$ is independent of a, except on those edges incident with v. Consequently, in the conditional probability

$$I(\omega)/\sum_{a \in A} I(\omega_a)$$

the product terms involving edges not incident with v may be cancelled throughout. The remaining terms depend only on the values of ω at v and its neighbours, and so I has the Markov property.

A partial converse of the preceding result is also true. If G satisfies some simple conditions, then given any weight function I on $\Omega(G, A)$ which has the Markov property, there is some interaction function i such that I is derived from i by the product formula (1.3.1). This converse enables us to use the interaction model in the study of stochastic processes on graphs. The equilibrium distributions of such processes often turn out to have the Markov property; consequently, they may be described in terms of our interaction model.

1.4 Physical background

An interaction model whose weight function is of the form $I(\omega) = \exp J(\omega)$ is typical of the models studied in statistical mechanics. The usual distribution of states is the Gibbs canonical distribution, wherein the weight of a state ω is equal to

$$\exp\{(-1/kT)H(\omega)\}. \qquad (1.4.1)$$

Here k is an absolute constant, T is the temperature, and $H(\omega)$ is a Hamiltonian function representing the energy of the state ω. In general, $H(\omega)$ will be a sum of terms corresponding to individual interactions, and so our product formula for I is obtained, as mentioned in the previous section. The various thermodynamic quantities, such as free energy and specific heat, may be derived from the partition function considered as a function of T.

For the purposes of illustration we shall discuss a famous example of an interaction model - the Ising model of ferromagnetism. In physical terms, the particles are the atoms of a ferromagnetic substance, each of which has a magnetic moment or 'spin'. There are just two possible configurations for each spin, and they are conventionally thought of as 'up' and 'down'. Like spins contribute an amount of energy -L to the Hamiltonian, and unlike spins contribute +L.

In our notation, the two configurations 'up' and 'down' become the two elements 0 and 1 of the ring A_2 of residues modulo 2. In accordance with the Gibbs distribution, the weight function is of the form

$$I(\omega) = \exp\{(-1/kT)\Sigma(\pm L)\} .$$

So if we put $\varepsilon = \exp(L/kT)$ and define an interaction function i on A_2 by the rules

$$i(0) = \varepsilon, \qquad i(1) = \varepsilon^{-1},$$

we have precisely the weight function of the **Ising model** $\mathcal{I} = (A_2, i)$, in terms of our definition (1.3.1). In fact, we have a set of models, one for each positive value of the temperature T.

It is fairly easy to find explicit expressions for the partition function of the Ising model on various well-known graphs. For example, consider the complete graph K_n, which has n vertices and $\frac{1}{2}n(n-1)$ edges, one edge joining each pair of distinct vertices; this graph corresponds to a physical system in which each pair of particles interacts. If the state ω on K_n has l up-spins and $n - l$ down-spins, then there are $l(n-l)$ unlike pairs and $\frac{1}{2}n(n-1) - l(n-l)$ like pairs; so the weight of ω is

$$I(\omega) = \varepsilon^{\frac{1}{2}n(n-1)-2l(n-l)}.$$

The partition function is

$$Z(\mathcal{I}, K_n) = \varepsilon^{\frac{1}{2}n(n-1)} \sum_{l=0}^{n} \binom{n}{l} \varepsilon^{-2l(n-l)}. \qquad (1.4.2)$$

There are a couple of fairly simple reduction formulae which can be employed to advantage. First, if G is a disconnected graph with (say) two components G_1 and G_2 then for any model $\mathfrak{M}$ we have

$$Z(\mathfrak{M}, G) = Z(\mathfrak{M}, G_1)Z(\mathfrak{M}, G_2). \qquad (1.4.3)$$

This may be proved directly from the definitions, just by noting that each state on G splits into a state on G_1 and a state on G_2, and there are no interactions between the two components. (See Theorem 7.) Secondly, we consider the process of 'deletion and contraction'. Given a graph G and an edge e of G (not a loop), let G' denote the graph obtained from G by deleting e, and G'' the graph obtained from G' by identifying the vertices which were the ends of e in G. For example, if G is the circuit graph C_n with n vertices (Fig. 3), then G' is the path graph P_n and G'' is the circuit graph C_{n-1}.

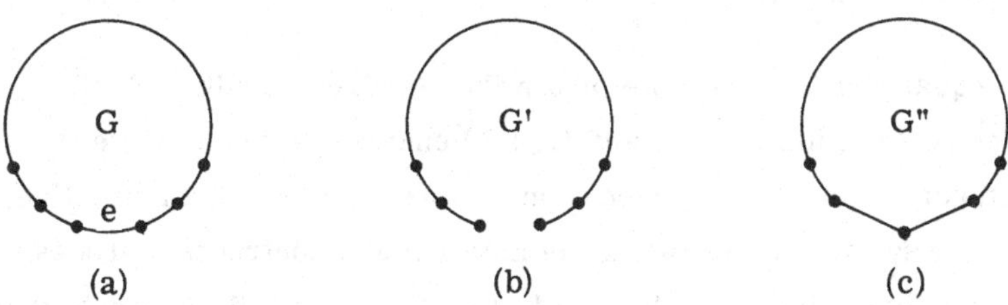

Fig. 3. Deletion and contraction

The partition functions for the Ising model on G, G', and G'' are related by the identity

$$Z(\mathcal{I}, G) = \varepsilon^{-1}Z(\mathcal{I}, G') + (\varepsilon - \varepsilon^{-1})Z(\mathcal{I}, G''). \tag{1.4.4}$$

To prove (1.4.4), let us denote the ends of e by u and v, and split the partition function for G into two parts:

$$Z(\mathcal{I}, G) = \Sigma_1(G) + \Sigma_2(G), \tag{1.4.5}$$

where Σ_1 is the sum over those states ω for which $\omega(u) = \omega(v)$, and Σ_2 is the sum over those for which $\omega(u) \neq \omega(v)$. Now the states contributing to $\Sigma_1(G)$ are in one-to-one correspondence with the states on G'', and the weight of such a state on G is ε times the weight of the corresponding state on G'', because of the presence of the edge e in G with like spins at its ends. Thus

$$\Sigma_1(G) = \varepsilon Z(\mathcal{I}, G''). \tag{1.4.6}$$

Applying the same arguments to G', we get

$$Z(\mathcal{I}, G') = \Sigma_1(G') + \Sigma_2(G') , \tag{1.4.7}$$

and

$$\Sigma_1(G') = Z(\mathcal{I}, G'') . \tag{1.4.8}$$

Also, since G' differs from G only in that e is absent, and this removes a factor ε^{-1} from each state contributing to $\Sigma_2(G)$, we have

$$\Sigma_2(G') = \varepsilon \, \Sigma_2(G) . \tag{1.4.9}$$

From the equations (1.4.5-9) we obtain the required result (1.4.4).

The two results (1.4.3) and (1.4.4) enable one to calculate the partition function for an Ising model on a 'small' graph, in an effective, but tedious, way, just by iterating the deletion and contraction process. We may also determine $Z(\mathcal{I}, P_n)$ and $Z(\mathcal{I}, C_n)$. For P_n, we note that P'_n has two components - an isolated vertex (P_1) and the graph P_{n-1} - provided that we choose the deleted edge to be one of the extreme ones. In this case, P''_n is just P_{n-1}. Thus

$$\begin{aligned} Z(\mathcal{I}, P_n) &= \varepsilon^{-1} Z(\mathcal{I}, P_{n-1}) Z(\mathcal{I}, P_1) + (\varepsilon - \varepsilon^{-1}) Z(\mathcal{I}, P_{n-1}) \\ &= (\varepsilon + \varepsilon^{-1}) Z(\mathcal{I}, P_{n-1}), \end{aligned}$$

where we have used the simple fact that $Z(\mathcal{I}, P_1) = 2$. Consequently,

$$Z(\mathcal{I}, P_n) = 2(\varepsilon + \varepsilon^{-1})^{n-1}. \tag{1.4.10}$$

In a similar way, using the reduction illustrated in Fig. 3, we find

$$Z(\mathcal{I}, C_n) = (\varepsilon + \varepsilon^{-1})^n + (\varepsilon - \varepsilon^{-1})^n. \tag{1.4.11}$$

We shall make use of this formula in the next section.

1.5 Transition points

The original aim of the Ising model was to explain certain aspects of ferromagnetism - in particular, the existence of a transition temperature T_c. At temperatures $T < T_c$ a ferromagnetic substance will exhibit some permanent magnetism, independent of an external field, while at temperatures $T > T_c$ there is no such permanent magnetisation. We have to explain how the partition function can have an isolated singularity at some point $T_c > 0$, although the interactions themselves are well-behaved for all $T > 0$.

We have already remarked that the Ising model is really a family of models, one for each $T > 0$. This idea will now be formulated at a level of generality suitable for the discussion of physical transitions.

A family of interaction models consists of a ring A and a function i defined on pairs (a, t), where a belongs to A and t is a positive real number. We shall write $i_a[t]$ for $i(a, t)$, and insist that for each a in A, i_a is an infinitely differentiable function of t, for $t > 0$. In addition, we make the usual symmetry requirement, $i_a = i_{-a}$. At each particular value of t we have an interaction model $\mathfrak{M}[t]$, comprising the ring A and the interaction function whose value for the configuration a is $i(a, t)$. The parameter t will usually be enclosed in square brackets, to avoid confusion with other variables. We shall use such abbreviations as $Z[t]$ for $Z(\mathfrak{M}[t], G)$, when $\mathfrak{M}$ and G are defined by the context. For example, the Ising family of models $\mathcal{I}[T]$ is defined by

$$A = A_2, \qquad i_0[T] = \exp(L/kT), \qquad i_1[T] = \exp(-L/kT),$$

and in the context of a discussion of the circuit graphs C_n we could use the notation

$$Z_n[T] = (2\cosh(L/kT))^n + (2\sinh(L/kT))^n.$$

As a preliminary suggestion, we might propose to define a 'transition point', for a family $\mathfrak{M}[t]$ and a graph G, to be a real number $t_c > 0$ such that $Z[t]$ or one of its derivatives does not exist at t_c, but does exist at all other points in some neighbourhood of t_c. It must be emphasised that the 'local' interactions are well-behaved for all real and positive values of t, but at a transition point the partition function (which determines the 'global' properties of the system) has an isolated singularity.

For finite graphs (the only case we have been considering!) such transition points cannot occur, since Z is the sum of a finite number of weights and each weight is a product of well-behaved interaction functions. In order to overcome this apparently basic defect of the mathematical formulation, it is necessary to consider 'large', or infinite, graphs. Certainly, physical systems contain a large number of particles, but not an infinite number, and so we must proceed with some care. This topic is one of the major themes of the book, and we shall begin here with a descriptive account, using the family of Ising models as an illustration.

Suppose that $\{G_n\}$ is a sequence of finite graphs with $|VG_n| = v_n$, and let $Z_n[T] = Z(\mathscr{I}[T], G_n)$. Then the 'thermodynamic limit'

$$Z_\infty[T] = \lim_{n\to\infty} \{Z_n[T]\}^{1/v_n}$$

may be regarded as the partition function for the 'limit' of the sequence $\{G_n\}$. Let $Z_n^{(l)}$ denote the lth derivative of Z_n $(l \geq 0)$. It might happen that, for some l, the behaviour of the sequence $\{Z_n^{(l)}[T]\}^{1/v_n}$ in the neighbourhood of T_c is very roughly like that shown in Fig. 4(a), so that the graph of $Z_\infty^{(l)}$ is as in (b).

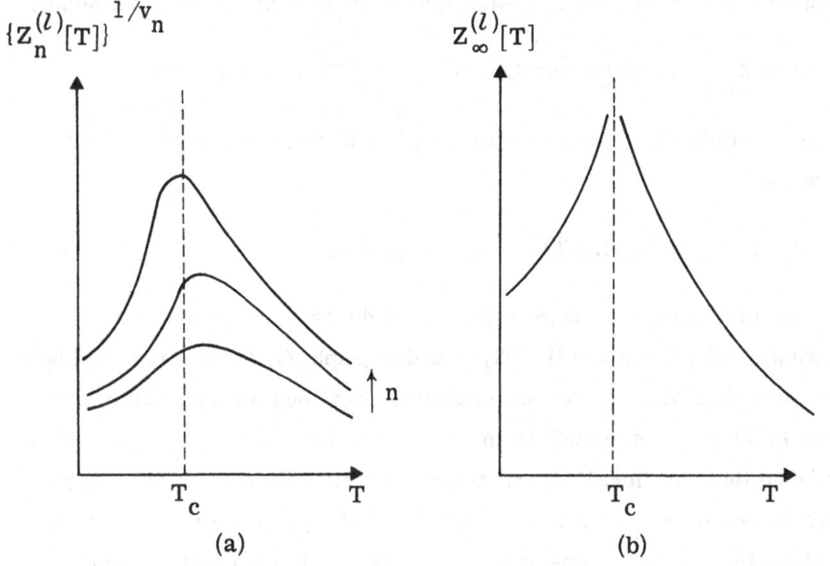

Fig. 4. A transition point

Such behaviour would indicate the existence of some experimentally observable phenomenon at the temperature T_c, for objects having a large number of particles, while T_c is a 'transition point' of Z_∞, in the sense proposed above. The singularity in the infinite limit manifests itself as a physical fact for large (but finite) systems.

In a few cases it is easy to find the exact behaviour of the limit. For the circuit graphs C_n the function

$$Z_\infty[T] = \lim_{n\to\infty} \{Z(\mathscr{I}[T], C_n)\}^{1/n}$$

may be regarded as the partition function for an infinite linear chain. From (1.4.11) with $\varepsilon = \exp(L/kT)$ we see that

$$Z_\infty[T] = 2\cosh(L/kT),$$

so that there are no transition points. This rather disappointing result was obtained by Ising in his original paper, and it led him to conclude (wrongly) that the model had no physical significance. For the sequence of complete graphs K_n, the limiting behaviour depends upon the sign of L. When L is negative we find

$$Z_\infty[T] = 2\exp(-L/2kT),$$

while when L is positive the limit does not exist for any $T > 0$. In neither case do we have a transition point.

However, there are some graphs which do show the desired kind of singularity. The two-dimensional analogue of the linear chain - the plane square lattice - has a partition function whose second derivative does not exist at a unique, real and positive, value T_c. This remarkable result is one of the main reasons for the importance of interaction models in theoretical physics, and it will be a constant source of motivation throughout this book.

We can now say precisely what is meant by the term 'transition point'. Let $\mathfrak{M}[t]$ be a family of interaction models, and $\{G_n\}$ a sequence of finite graphs, with $|VG_n| = v_n$. Define

$$Z_\infty[t] = \lim_{n\to\infty} \{Z(\mathfrak{M}[t], G_n)\}^{1/v_n}.$$

The real and positive number t_c is a **transition point** for $\mathfrak{M}[t]$ on $\{G_n\}$ if $Z_\infty[t]$, or one of its derivatives, exists throughout some neighbourhood of t_c, but does not exist at t_c itself.

We remark that the definition applies, not to a single graph (finite or infinite), but to a sequence of finite graphs. It might be thought desirable to define the partition function of an infinite graph first by means of the thermodynamic limit, and then to formulate the notion of a transition point in terms of that definition. But in order to follow such a procedure we should have to face very considerable problems of existence and uniqueness at the outset: for example, the infinite linear chain may be regarded as the limit of a sequence of finite graphs in many different ways. Nevertheless, we shall often overlook such difficulties, and speak of 'the partition function of the infinite graph G', when what we really mean is 'the limit of a sequence $\{Z(G_n)\}^{1/v_n}$ for a sequence $\{G_n\}$ of finite graphs which tends (in some sense) to G'.

It may be appropriate to conclude this introductory chapter by mentioning some general points which, for reasons of space, will not be discussed in the remainder of the book.

There are now very many recorded examples of critical phenomena, analogous to the existence of a transition temperature for a ferromagnetic substance. Perhaps the most basic concerns the liquid-gas transition (condensation), and there are others associated with superfluidity, superconductivity, and ferroelectricity. In all cases the mathematical problem is essentially the same: namely, to account for singularities at a macroscopic level in terms of microscopic interactions which are, presumably, well-behaved.

The simple mathematical model which we propose to study in this book is clearly capable of several instant generalizations. We may allow the set of configurations to be countably or uncountably infinite, and we may introduce interactions between more than two particles. The justification for discussing the simplest kind of model is twofold. First, such models do in fact exhibit transition points of the kind observed in experimental physics, and secondly, there is still much work to be done before even these models are fully understood.

NOTES AND REFERENCES FOR CHAPTER 1

For background reading in graph theory the reader is referred to the introductory text by Wilson [9], and the present author's Cambridge Tract [1].

The probabilistic treatment of interaction models is surveyed by Preston [7]. His book contains a neat proof of the results on the Markov property for finite graphs, due to Grimmett [4].

A readable introduction to statistical mechanics is the book by Thompson [8], which contains a great deal of material relevant to the topics mentioned in Sections 1.3-1.5. The history of the Ising model is traced in an interesting article by Brush [2]. The work of Peierls [6] and Onsager [5], on the critical temperature of the plane square lattice, deserves special mention. The whole field of critical phenomena is surveyed in a series of volumes edited by Domb and Green [3].

1. Biggs, N. L. Algebraic graph theory. (Cambridge University Press, 1974.)
2. Brush, S. G. History of the Lenz-Ising model. Rev. Mod. Phys. 39 (1967), 883-93.
3. Domb, C. and Green, M. S. (eds.). Phase transitions and critical phenomena, Volumes 1-6. (Academic Press, New York, 1972-6.)
4. Grimmett, G. R. A theorem about random fields. Bull. London Math. Soc., 5 (1973), 81-4.
5. Onsager, L. Crystal statistics, I. A two-dimensional model with an order-disorder transition. Phys. Rev., 65 (1944), 117-49.
6. Peierls, R. Ising's model of ferromagnetism. Proc. Cambridge Philos. Soc., 32 (1936), 477-81.
7. Preston, C. Gibbs states on countable sets. (Cambridge University Press, 1974.)
8. Thompson, C. J. Mathematical statistical mechanics. (Macmillan, New York, 1972.)
9. Wilson, R. J. Introduction to graph theory. (Oliver and Boyd, Edinburgh, 1972.)

2 · Methods

2.1 Resonant models

In this chapter we shall develop some of the mathematical apparatus used in the discussion of interaction models and their transition points. We begin with a simplified kind of model, which has three useful properties: it provides easily-worked examples, it exhibits typical kinds of singular behaviour, and it subsumes most of the models studied in practice.

There are some circumstances in which it is reasonable to assume that the interaction which takes place between two particles depends only on whether or not they have the same configuration. For example, it might happen that the interaction exhibits some kind of 'resonance' which favours like configurations. When this is the case, our general model $\mathfrak{M} = (A, i)$ has an interaction function of the form

$$i(0) = i_0, \qquad i(a) = i_1 \qquad (a \neq 0). \tag{2.1.1}$$

We shall say that such a model is **resonant**, and use the general symbol $\mathcal{R}$ (or $\mathcal{R}_i$ if i is specified) for models of this kind. We shall let m denote the number of possible configurations for each particle: $m = |A|$.

Examples of resonant interaction models have already appeared in Chapter 1; for instance, the colouring model $\mathcal{C}$ defined in Section 1.3 is of the resonant kind. Of course, when each particle has only two possible configurations $(m = 2)$ the model is necessarily resonant, so the Ising model is an important example. The resonant model with $i_0 = \varepsilon$, $i_1 = \varepsilon^{-1}$ and $m > 2$ is known in physics as the **Potts model**.

There are two very simple cases where it is possible to give an explicit formula for $Z(\mathcal{R}, G)$. These cases will be useful in the ensuing theory.

First, let $m = 2$ and h denote the interaction function defined on A_2 by

$$h(0) = 1, \quad h(1) = -1. \tag{2.1.2}$$

It will be convenient to consider the elements 0 and 1 of A_2 as real numbers, so that we may write

$$h(a) = (-1)^a \qquad (a \in A_2).$$

Then for any state ω in $\Omega(G, A_2)$, and any e in EG,

$$\begin{aligned} h\{\delta\omega(e)\} &= h\{\omega(x) - \omega(y)\} \\ &= (-1)^{\omega(x)-\omega(y)} \\ &= (-1)^{\omega(x)}(-1)^{\omega(y)}, \end{aligned}$$

where x and y are the ends of e. The weight of ω is the product of terms $(-1)^{\omega(v)}$, each vertex v contributing a number of times equal to its valency $k(v)$ (the number of edges incident with v). Consequently, the partition function is

$$Z(\mathcal{R}_h, G) = \sum_{\omega: V \to A_2} \prod_{v \in V} (-1)^{\omega(v)k(v)}.$$

Applying Lemma I of Appendix A, we may rewrite this as

$$\prod_{v \in V} \sum_{a \in A_2} (-1)^{ak(v)}.$$

Since A_2 has only two elements 0 and 1,

$$\sum_{a \in A_2} (-1)^{ak(v)} = \begin{cases} 2 & \text{if } k(v) \text{ is even;} \\ 0 & \text{if } k(v) \text{ is odd.} \end{cases}$$

Hence,

$$Z(\mathcal{R}_h, G) = \begin{cases} 2^{|VG|} & \text{if every vertex of } G \text{ has even valency;} \\ 0 & \text{otherwise.} \end{cases} \tag{2.1.3}$$

Connected graphs in which each vertex has even valency are familiar to every student who has taken an introductory course in graph theory. Such graphs have the Eulerian property: they may be covered by a closed path which traverses each edge just once.

Now suppose $m \geq 2$ and let f be the interaction function

$$f(0) = f_0, \quad f(a) = 0 \qquad (a \neq 0). \tag{2.1.4}$$

The weight

$$I(\omega) = \Pi f\{\delta\omega(e)\}$$

is zero if $\delta\omega(e) \neq 0$ for any edge e of G. Consequently, the only states which make a non-zero contribution to the partition function are those which assign the same element of A to each pair of vertices joined by an edge: that is, those which are constant on each component of G. If G has $c(G)$ components, then there are $m^{c(G)}$ such states, and each one contributes a weight $f_0^{|EG|}$. Thus

$$Z(\mathcal{R}_f, G) = m^{c(G)} f_0^{|EG|}. \tag{2.1.5}$$

In practice, we should not expect that the values of a resonant model will fit neatly into the two simple cases just described. However, an elementary trick enables us to express a partition function $Z(\mathcal{R}, G)$ in terms of the partition function of some other resonant model $\mathcal{R}'$, evaluated on subgraphs of G. We remark first that any pair of interaction functions, g and i, defined on the same ring A and having the form of (2.1.1) are related by a transformation

$$i(a) = pg(a) + q \qquad (a \in A),$$

for some complex numbers p and q. Explicitly, if the values of g are g_0 and g_1 and those of i are i_0 and i_1, then (provided $g_0 \neq g_1$)

$$p = \frac{i_0 - i_1}{g_0 - g_1}, \qquad q = \frac{i_1 g_0 - i_0 g_1}{g_0 - g_1}.$$

When g and i are so related, the weight function corresponding to i is

$$I(\omega) = \prod_{e \in E} \{pg[\delta\omega(e)] + q\}.$$

On multiplying out the terms in the product, we obtain a summand corresponding to each subset S of E: it is the product of $|E - S|$ factors q and the factors $pg[\delta\omega(e)]$ for e in S. Consequently,

$$I(\omega) = \sum_{S\subseteq E} q^{|E-S|} p^{|S|} \prod_{e\in S} g[\delta\omega(e)]$$

$$= q^{|E|} \sum_{S\subseteq E} (p/q)^{|S|} \prod_{e\in S} g[\delta\omega(e)]. \qquad (2.1.6)$$

The transformation leading to (2.1.6) occurs in many disguises: it is famous in theoretical physics because of its use by Mayer in his theory of condensation in 1937. From our viewpoint, it may be used to generate subgraph expansions at will, as shown in Theorem 1 below.

For each subset S of the edge-set E of a graph G, we define the edge-subgraph $\langle S\rangle_G$, generated by S, to be the graph formed by the edges in S and those vertices of G which are incident with some edge in S. We often write $\langle S\rangle$, instead of $\langle S\rangle_G$. It will also be convenient to use the abbreviation $\overline{Z}$ for $Z/|\Omega|$, that is

$$\overline{Z}(\mathfrak{M}, G) = \frac{Z(\mathfrak{M}, G)}{m^{|VG|}} .$$

Theorem 1. <u>Let</u> $\mathcal{R}_i = (A, i)$ <u>and</u> $\mathcal{R}_g = (A, g)$ <u>be resonant interaction models with</u>

$$i(a) = pg(a) + q \qquad (a \in A).$$

<u>Then for any graph</u> G, <u>we have</u>

$$\overline{Z}(\mathcal{R}_i, G) = q^{|E|} \sum_{S\subseteq E} \left(\frac{p}{q}\right)^{|S|} \overline{Z}(\mathcal{R}_g, \langle S\rangle). \qquad (2.1.7)$$

Proof. From the definition of $Z(\mathcal{R}_i, G)$ and the expression (2.1.6) for $I(\omega)$, we obtain

$$Z(\mathcal{R}_i, G) = q^{|E|} \sum_{\omega\in\Omega} \sum_{S\subseteq E} (p/q)^{|S|} \prod_{e\in S} g[\delta\omega(e)],$$

where $\Omega = \Omega(G, A)$. Interchanging the order of summation yields

$$Z(\mathcal{R}_i, G) = q^{|E|} \sum_{S\subseteq E} (p/q)^{|S|} \sum_{\omega\in\Omega} \prod_{e\in S} g[\delta\omega(e)].$$

Now each state ω on G induces by restriction a state on $<S>$, and each state on $<S>$ is the restriction of $m^{|VG| - |V<S>|}$ states on G. Hence

$$\sum_{\omega \in \Omega} \prod_{e \in S} g[\delta\omega(e)] = m^{|VG| - |V<S>|} Z(\mathcal{R}_g, <S>),$$

which gives the required result. //

Theorem 1 has many applications. The Ising model $\mathcal{I}$ is clearly ripe for comparison with the simple model $\mathcal{R}_h$ whose partition function is given by (2.1.3). We have

$$i_0 = \varepsilon, \quad i_1 = \varepsilon^{-1}, \quad h_0 = 1, \quad h_1 = -1,$$

so that $p = \frac{1}{2}(\varepsilon - \varepsilon^{-1})$ and $q = \frac{1}{2}(\varepsilon + \varepsilon^{-1})$. In terms of the variable $\nu = (L/kT)$, we have $\varepsilon = \exp \nu$, $p = \sinh \nu$, and $q = \cosh \nu$. Thus, by Theorem 1,

$$\overline{Z}(\mathcal{I}, G) = (\cosh \nu)^{|E|} \sum_{S \subseteq E} \overline{Z}(\mathcal{R}_h, <S>) \, (\tanh \nu)^{|S|}.$$

From (2.1.3) we see that the coefficient of $(\tanh \nu)^{|S|}$ is zero if any vertex of $<S>$ has odd valency, and 1 otherwise. Hence

$$\overline{Z}(\mathcal{I}, G) = (\cosh \nu)^{|E|} \sum_{l} N(l) \, (\tanh \nu)^{l}, \qquad (2.1.8)$$

where $N(l)$ is the number of edge-subgraphs of G which are 'admissible' (that is, which have the property that each vertex has even valency) and have l edges.

The formula (2.1.8) may be regarded as a series expansion for the partition function of the Ising model. If the temperature T is large, then $\tanh \nu$ is small, and the series is consequently known as a 'high-temperature expansion' for the Ising model. The set of admissible subgraphs is small compared with the set of all subgraphs, and so the method is quite efficient in some cases. For example, consider the path graphs and circuit graphs; a path graph has just one admissible subgraph, corresponding to $S = \emptyset$, and a circuit graph has two, $S = \emptyset$ and $S = E$. The equations (1.4.10) and (1.4.11) may thus be verified at once. For the graph $K_{3,3}$ depicted in Fig. 5, we find the values of $N(l)$ shown in the table, from which the partition function may be written down, if required.

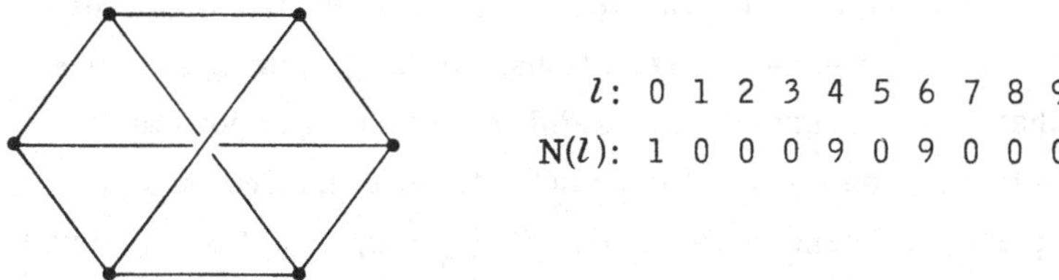

l:	0	1	2	3	4	5	6	7	8	9
$N(l)$:	1	0	0	0	9	0	9	0	0	0

Fig. 5. Admissible subgraphs of $K_{3,3}$

We shall derive the high-temperature expansion in a quite different way in the next chapter.

Another useful application of Theorem 1 arises from the comparison of a general resonant model $\mathcal{R}_i$ with the special model $\mathcal{R}_f$ defined by (2.1.4). Putting $g_0 = f_0$ and $g_1 = 0$ in the equations for p and q, we get

$$p = (i_0 - i_1)/f_0, \quad q = i_1.$$

So, using the abbreviation $|VS|$ for $|V\langle S\rangle|$, the theorem states that

$$\overline{Z}(\mathcal{R}_i, G) = i_1^{|E|} \sum_{S\subseteq E} \left(\frac{i_0 - i_1}{f_0 i_1}\right)^{|S|} \frac{m^{c\langle S\rangle} f_0^{|S|}}{m^{|VS|}}$$

$$= i_1^{|E|} \sum_{S\subseteq E} (i_0/i_1 - 1)^{|S|} m^{c\langle S\rangle - |VS|}.$$

For any graph H, the number $r(H) = |VH| - c(H)$ is called the **rank** of H; it is the rank of the incidence matrix of H as defined in Section 1.3. With this notation we have shown that

$$\overline{Z}(\mathcal{R}_i, G) = i_1^{|E|} \sum_{S\subseteq E} (i_0/i_1 - 1)^{|S|} \left(\frac{1}{m}\right)^{r\langle S\rangle}. \qquad (2.1.9)$$

The right-hand-side of (2.1.9) is an example of the 'rank polynomial' of a graph. The fact that the partition function of a resonant model can be expressed in this way has two important implications.

First, we see that Z depends essentially only on the ratio i_0/i_1; this is fairly obvious from the definition. Secondly, we notice that, in the resonant case, Z depends only on the number m of configurations available to each particle, not on any algebraic structure we might have imposed on the set A of such configurations. In the following sections we shall see that it is, nevertheless, useful to be able to impose an algebraic structure, even though it does not influence the final result.

In the particular case of the model $\mathcal{C}$ ($i_0 = 0$, $i_1 = 1$), it is usual to treat m as a variable. The number of proper colourings is given by

$$Z(\mathcal{C}, G) = m^{|VG|} \sum_{l} C_l(G)\left(\frac{1}{m}\right)^l , \qquad (2.1.10)$$

where the coefficients are

$$C_l(G) = \sum (-1)^{|S|} ,$$

the sum being taken over all edge-subgraphs $\langle S \rangle$ of G whose rank is l. Incidentally, we note that $Z(\mathcal{C}, G)$ is a polynomial in m, of degree $|VG|$, so that the name 'chromatic polynomial' is justified. The dominant coefficients are those for small values of l, and these depend only on 'small' subgraphs. For instance, the only edge-subgraph of rank 0 is the null graph, and the only ones of rank 1 are the single edges, so

$$C_0(G) = 1, \quad C_1(G) = -|EG| .$$

The resonant models discussed in this section are a very special class of interaction models. In a later chapter we shall find that more general models also admit subgraph expansions, and that the general theory leads to much stronger results. In the remainder of this chapter we shall return to our main theme, and introduce an algebraic technique which throws some light on the question of transition points.

2.2 The transfer matrix

Exact calculations of the partition function are often based on the 'transfer matrix' method. We begin with a method of constructing a sequence of finite graphs, each having cyclic symmetry. The construction will be introduced in a fairly down-to-earth way, but later in the book we shall relate it to more fundamental questions of symmetry and dimensionality.

A graph scheme G is a pair (F, J), where F is a graph and J is a subset of the set $VF \times VF$ of ordered pairs of vertices of F. For a given graph scheme G and any integer $n \geq 3$, we may construct a graph G_n, in the manner depicted in Fig. 6. To be precise, G_n is the

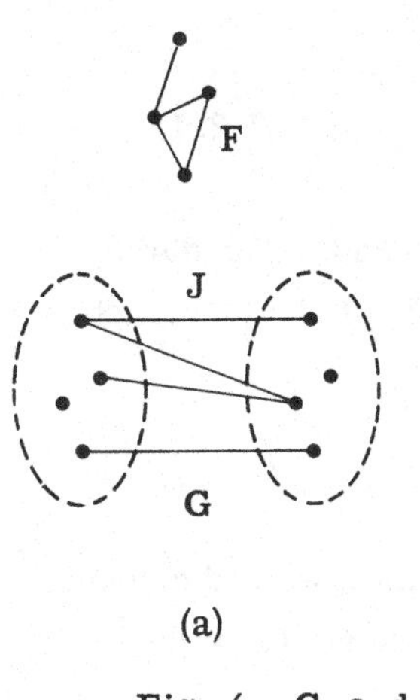

(a)

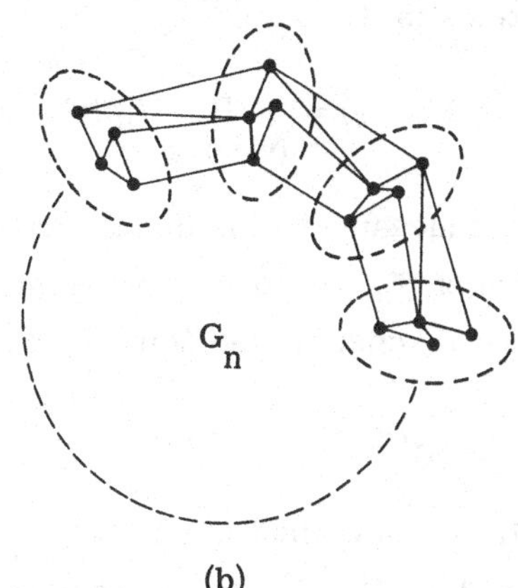

(b)

Fig. 6. G and G_n

graph whose vertex-set is the union

$$V_n = V^{(1)} \cup V^{(2)} \cup \ldots \cup V^{(n)},$$

and whose edge-set is the union

$$E_n = E^{(1)} \cup E^{(2)} \cup \ldots \cup E^{(n)},$$

where $V^{(l)}$ is a set of vertices v_l, one for each v in VF, and $E^{(l)}$ is a set of edges, called e_l and j_l, defined as follows. There is one edge e_l for each e in EF and each $l = 1, 2, \ldots, n$, and if e is incident with the vertices v and w in F, then e_l is incident with the vertices v_l and w_l. Similarly, there is one edge j_l for each j in J and each $l = 1, 2, \ldots, n$, and if $j = (x, y)$ then j_l is incident with x_l

and y_{l+1}, where $n + 1$ is taken to be the same as 1. Roughly speaking, G_n is formed by taking n copies of the graph F and joining them cyclically with linking edges as prescribed by the set J. The simplest example is the scheme C, with F consisting of a single vertex v and J the pair (v, v); the resulting graph C_n is, of course, the familiar circuit graph.

Let $\mathfrak{M} = (A, i)$ be an interaction model, and G a graph scheme. Write Ω for $\Omega(F, A)$ and Ω_n for $\Omega(G_n, A)$. For each pair (σ, π) of states in Ω, define

$$L(\sigma, \pi) = \prod_{(v, w) \in J} i[\sigma(v) - \pi(w)]. \tag{2.2.1}$$

L is a measure of the interaction between states σ and π on disjoint copies of F, due to the linking edges specified by J. Let $I(\sigma)$ denote the usual weight of the state σ on F, and put

$$T_G(\sigma, \pi) = I(\sigma)\, L(\sigma, \pi). \tag{2.2.2}$$

It will be convenient to think of T_G as a matrix, with rows and columns indexed by Ω; T_G is the **transfer matrix** of the scheme G with respect to the model $\mathfrak{M}$.

Theorem 2. <u>Let $T = T_G$ be the transfer matrix of a graph scheme G with respect to an interaction model $\mathfrak{M}$. The partition function of $\mathfrak{M}$ on G_n is given by</u>

$$Z(\mathfrak{M}, G_n) = \text{trace } T^n. \tag{2.2.3}$$

Proof. There is a one-to-one correspondence between the states ω on G_n and the ordered n-tuples $(\omega_1, \omega_2, \ldots, \omega_n)$, where each ω_l represents the state on F resulting from the restriction of ω to the subset $V^{(l)}$ of V_n. The weight of ω on G_n is the product of factors $i[\delta\omega(q)]$, one for each edge q of G_n. These edges are of two kinds: the edges e_l for which the corresponding factor is $i[\delta\omega_l(e_l)]$, and the edges j_l for which the corresponding factor is

$$i[\omega_l(x_l) - \omega_{l+1}(y_{l+1})],$$

when $j = (x, y)$. Hence the weight of ω is

$$\prod_{l=1}^{n} I(\omega_l)L(\omega_l, \omega_{l+1}) = \prod_{l=1}^{n} T(\omega_l, \omega_{l+1}).$$

Consequently,

$$\begin{aligned} Z(\mathfrak{M}, G_n) &= \sum_{\omega\in\Omega_n} \prod_{l=1}^{n} T(\omega_l, \omega_{l+1}) \\ &= \sum_{(\omega_1, \omega_2, \ldots, \omega_n)} \prod_{l} T(\omega_l, \omega_{l+1}) \\ &= \sum_{\omega_1} \sum_{(\omega_2, \ldots, \omega_n)} T(\omega_1, \omega_2)T(\omega_2, \omega_3)\ldots T(\omega_n, \omega_1) \\ &= \sum_{\omega_1} T^n(\omega_1, \omega_1) \\ &= \text{trace } T^n. \ // \end{aligned}$$

2.3 Applications of the trace formula

The usefulness of Theorem 2 stems from the fact that the trace of T^n may be expressed in terms of the spectrum of T. If $\lambda_1, \lambda_2, \ldots, \lambda_r$ $(r = |\Omega|)$ are the eigenvalues of T, each taken with the correct (algebraic) multiplicity, then the eigenvalues of T^n are $\lambda_1^n, \lambda_2^n, \ldots, \lambda_r^n$, and

$$\text{trace } T^n = \lambda_1^n + \lambda_2^n + \ldots + \lambda_r^n. \qquad (2.3.1)$$

For example, suppose C is the circuit graph scheme and $\mathfrak{M} = (A, i)$ is any model with $A = A_m$, a ring of residues. The transfer matrix T is the circulant

$$\begin{bmatrix} i(0) & i(-1) & \ldots & i(1) \\ i(1) & i(0) & \ldots & i(2) \\ \cdot & \cdot & & \cdot \\ \cdot & \cdot & & \cdot \\ \cdot & \cdot & & \cdot \\ i(-1) & i(-2) & \ldots & i(0) \end{bmatrix}$$

whose eigenvalues are

$$\lambda_j = i(0) + i(1)\eta_j + \dots + i(m-1)\eta_j^{m-1}$$

where $\eta_j = \exp(2\pi ij/m)$ and j runs from 1 to m. So in this case we may write down an explicit formula for $Z(\mathfrak{M}, C_n)$. We note that the symmetry of the interaction, $i(a) = i(-a)$, ensures that all eigenvalues are real. For a resonant model $\mathfrak{R}$, we have

$$i(0) = i_0, \quad i(1) = i(2) = \dots = i(m-1) = i_1,$$

and

$$\lambda_j = i_0 - i_1 \qquad\qquad (1 \le j \le m-1)$$
$$\lambda_m = i_0 + (m-1)i_1.$$

Consequently,

$$Z(\mathfrak{R}, C_n) = (m-1)(i_0 - i_1)^n + (i_0 + (m-1)i_1)^n. \tag{2.3.2}$$

Special cases of (2.3.2) are the partition function of the Ising model (1.4.11), and the well-known expression for the chromatic polynomial of a circuit graph.

Of course, it is only rarely possible to find explicit formulae for the eigenvalues of the transfer matrix. Nevertheless, the method has considerable scope - in particular, as a tool in the study of transition points. We pass on to a discussion of this topic.

Let us suppose that the eigenvalues of T are arranged in order of their moduli:

$$|\lambda_1| \ge |\lambda_2| \ge |\lambda_3| \ge \dots \ge |\lambda_r|,$$

and that those of maximum modulus, $\lambda_1, \dots, \lambda_p$ say, are all in fact equal to the same value λ_{max}. In other words, λ_{max} is an eigenvalue of multiplicity p and all other eigenvalues have moduli strictly less than $|\lambda_{max}|$. Then applying (2.2.3) and (2.3.1) we have

$$Z(\mathfrak{M}, G_n) = \lambda_{max}^n [p + \sum_{l>p} (\lambda_l/\lambda_{max})^n].$$

Since $|\lambda_l / \lambda_{max}| < 1$ for $l > p$, and $|VG_n| = n|VF|$, we deduce that

$$\lim_{n \to \infty} \{Z(\mathfrak{M}, G_n)\}^{1/|VG_n|} = \lambda_{max}^{1/|VF|}. \qquad (2.3.3)$$

It is worth remarking that this result holds without restriction on the multiplicity of the eigenvalue λ_{max}; all that is required is that there should be no other eigenvalues with the same modulus.

The equation (2.3.3) is a fairly explicit indication of the existence of transition points, at least in the mathematical sense. If we have a family of models $\mathfrak{M}[t] = (A, i[t])$ and a given graph scheme G, then each individual eigenvalue of the transfer matrix will be a well-behaved function of t. However, the maximum of these functions will almost certainly be non-differentiable at the points where a new eigenvalue 'takes over' as the largest. An artificial example will make this clear. Suppose that we have a family of resonant models $\mathcal{R}[t]$, with interactions given by

$$i_0[t] = 1, \quad i_1[t] = t - 1.$$

The transfer matrix $T[t]$ of the circuit graph scheme C with respect to the model $\mathcal{R}[t]$ is

$$\begin{bmatrix} 1 & t-1 \\ t-1 & 1 \end{bmatrix},$$

which has eigenvalues $\lambda_1[t] = t$, $\lambda_2[t] = 2 - t$. The maximum eigenvalue $\lambda_{max}[t]$ is not differentiable at the point $t = 1$, where $\lambda_1[t] = \lambda_2[t]$. Thus the point $t = 1$ is a transition point for the family $\mathcal{R}[t]$ on the sequence $\{C_n\}$ of circuit graphs. We might say that $\mathcal{R}[t]$ exhibits a transition on the infinite linear chain (Fig. 7 overleaf).

It is a simple matter to construct similar examples of families of models with transition points. Nevertheless, it is possible to show that in 'physical' cases, when the models are positive, such transitions cannot occur within the framework provided by graph schemes. The underlying mathematical fact is the Perron-Frobenius theorem on the eigenvalues of positive matrices. For convenience, this result is stated and proved in Appendix B.

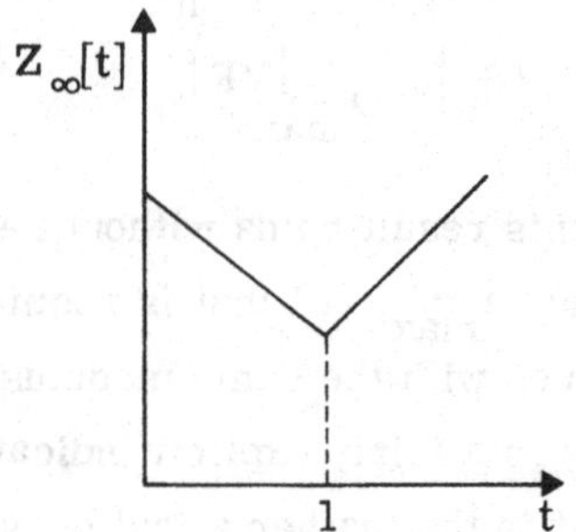

Fig. 7. An artificial transition

Theorem 3. Let $\mathcal{P}[t] = (A, i[t])$ be a family of positive models, and let G be any graph scheme. Then there are no transition points for the family $\mathcal{P}[t]$ on the sequence $\{G_n\}$; in other words, the limit

$$Z_\infty[t] = \lim_{n\to\infty} Z(\mathcal{P}[t], G_n)^{1/|VG_n|}$$

exists and is infinitely differentiable for all $t > 0$.

Proof. Since $\mathcal{P}[t]$ is a positive model, it follows that every entry of the relevant transfer matrix $T[t]$, as defined in (2.2.2), must be positive. Hence the theorem of Appendix B applies: for each $t > 0$ there is an eigenvalue $\lambda_{max}[t]$ which is real, positive, of maximum modulus, and it has unit multiplicity.

Furthermore, our definition of a family of models requires that the interactions are infinitely differentiable functions of t, and so the same is true of the entries of $T[t]$ and, locally, of its eigenvalues. In order that t_c should be a transition point, it is therefore necessary that at least two eigenvalues should be equal to λ_{max} at t_c, and this contradicts the Frobenius theorem. We conclude that there are no transition points. //

The positivity condition in Theorem 3 can be relaxed to a non-negativity condition, provided that some other conditions on the entries of the transfer matrix are satisfied. Such cases are important, for example when dealing with the colouring model $\mathcal{C}$, but it would be difficult to formulate a strengthened form of Theorem 3 which would cover all the cases which might arise.

2.4 Correlation functions

In theoretical physics there is an alternative approach to the problem of defining a phase transition, which uses the idea of 'order' instead of the basic partition function. Roughly speaking, we propose to examine the expected correlation between the configurations of particles separated by a large distance. The existence of a non-zero correlation indicates a degree of 'order' in the system, which should manifest itself as an observable physical fact. An obvious example occurs in the theory of magnetism, where the alignment of the individual magnetic moments (below the critical temperature) leads to the phenomenon of permanent magnetisation. We shall investigate the mathematical aspects of this idea, using the tools developed in the preceding sections.

Let $G = (F, J)$ be a graph scheme. Choose any two vertices x and y of F; these will be fixed throughout the discussion. The kth **correlation function** on G_n $(n > k)$ is the function c_k defined on states ω in $\Omega_n = \Omega(G_n, A)$ by

$$c_k(\omega) = \begin{cases} m - 1 & \text{if } \omega(x_1) = \omega(y_{k+1}); \\ -1 & \text{otherwise.} \end{cases}$$

The values are chosen so that, if all states were equally weighted, the expected value of c_k would be zero:

$$\sum_{\omega \in \Omega_n} c_k(\omega) = 0.$$

However, each state ω on G_n has a weight $I_n(\omega)$, and so we have to investigate the quantity

$$< c_k >_n = (1/Z) \sum_{\omega \in \Omega_n} c_k(\omega) I_n(\omega). \qquad (2.4.1)$$

The kth **correlation function** of the graph scheme G may be defined as

$$\gamma_k(G) = \lim_{n\to\infty} \langle c_k \rangle_n ,$$

and the order parameter of G (often called the 'long-range order') is

$$\gamma_\infty(G) = \lim_{k\to\infty} \gamma_k(G) = \lim_{k\to\infty} \lim_{n\to\infty} \langle c_k \rangle_n . \qquad (2.4.2)$$

As in the case of the partition function, the transfer matrix method facilitates the calculation of correlation functions and the order parameter. For each pair of states σ and τ in $\Omega = \Omega(F, A)$ we put

$$C(\sigma, \tau) = \begin{cases} m - 1 & \text{if } \sigma(x) = \tau(y); \\ -1 & \text{otherwise.} \end{cases}$$

C is regarded as a matrix, with rows and columns indexed in the same way as in the transfer matrix T. Using the one-to-one correspondence between states ω on G_n and n-tuples $(\omega_1, \omega_2, \ldots, \omega_n)$ of states on F, we have $c_k(\omega) = C(\omega_1, \omega_{k+1})$ and

$$\langle c_k \rangle_n = (1/Z) \sum_{(\omega_1, \ldots, \omega_n)} C(\omega_1, \omega_{k+1}) T(\omega_1, \omega_2) \ldots T(\omega_n, \omega_1).$$

Summing over all ω_i except ω_1 and ω_{k+1}, and renaming the dummy variables $\omega_1 = \sigma$, $\omega_{k+1} = \tau$, we obtain

$$\langle c_k \rangle_n = (1/Z) \sum_{\sigma, \tau} C(\sigma, \tau) T^k(\sigma, \tau) T^{n-k}(\tau, \sigma). \qquad (2.4.3)$$

In order to follow the method used for the partition function, we must make some restrictive hypothesis concerning the spectrum of T. We suppose that

$$T = \sum_{i=1}^{r} \lambda_i E_i , \qquad (2.4.4)$$

where $r = |\Omega|$, $\lambda_1, \ldots, \lambda_r$ are the eigenvalues of T (each occurring with its algebraic multiplicity) and $E_1, \ldots, E_r$ are mutually orthogonal idempotent matrices. That is

$$E_i E_j = 0 \qquad (i \neq j), \qquad E_i^2 = E_i .$$

The existence of this spectral decomposition of T is equivalent to the requirement that each eigenvalue has an eigenspace of maximal dimension, so that its 'geometric' multiplicity is equal to its algebraic multiplicity.

It follows from (2.4.3) that $T^s = \Sigma\lambda_i^s E_i$. We substitute in (2.4.4) and interchange the order of summation, obtaining

$$\langle c_k \rangle_n = (1/Z) \sum_{i,j} a_{ij}\lambda_i^k \lambda_j^{n-k}, \qquad (2.4.5)$$

where

$$a_{ij} = \sum_{\sigma,\tau} C(\sigma,\tau)\, E_i(\sigma,\tau)\, E_j(\tau,\sigma).$$

The order parameter is obtained from (2.4.5) by letting n, and then k, tend to infinity. The result obtained is influenced by two things: coincidences between the eigenvalues, and the possibility that some coefficients a_{ij} may be zero. These features are compatible with the existence of points of discontinuity in the order parameter of a family.

For example, suppose $A = A_2$ and C is the circuit graph scheme. As we found in Section 2.3, T is the matrix

$$\begin{bmatrix} i_0 & i_1 \\ i_1 & i_0 \end{bmatrix}$$

with eigenvalues $\lambda_1 = i_0 + i_1$ and $\lambda_2 = i_0 - i_1$. The spectral decomposition of T is $\lambda_1 E_1 + \lambda_2 E_2$, where

$$E_1 = \tfrac{1}{2}\begin{bmatrix} 1 & 1 \\ 1 & 1 \end{bmatrix}, \qquad E_2 = \tfrac{1}{2}\begin{bmatrix} 1 & -1 \\ -1 & 1 \end{bmatrix},$$

and the correlation matrix C is

$$\begin{bmatrix} 1 & -1 \\ -1 & 1 \end{bmatrix}.$$

It follows that $a_{11} = a_{22} = 0$ and $a_{12} = a_{21} = 1$, so that

$$\langle c_k \rangle_n = \frac{\lambda_1^k \lambda_2^{n-k} + \lambda_2^k \lambda_1^{n-k}}{\lambda_1^n + \lambda_2^n}.$$

To evaluate the order parameter when $|\lambda_1| > |\lambda_2|$, we divide through by λ_1^n and obtain

$$\gamma_k(C) = \lim_{n\to\infty} \langle c_k \rangle_n = (\lambda_2/\lambda_1)^k,$$

$$\gamma_\infty(C) = \lim_{k\to\infty} \gamma_k(C) = 0 .$$

Similarly, the order parameter is zero when $|\lambda_1| < |\lambda_2|$. But if $\lambda_1 = \lambda_2$, then

$$\gamma_k(C) = 1 \text{ and } \gamma_\infty(C) = 1.$$

We deduce that, in the case of a family of interaction models, there may be isolated points for which the long-range order is non-zero. For instance, when $i_0[t] = 1$ and $i_1[t] = t - 1$ (the example introduced in Section 2. 3), we have

$$\gamma_\infty[t] = \begin{cases} 0 & (t \neq 1), \\ 1 & (t = 1). \end{cases}$$

The discontinuity occurs at the point $t = 1$ where the partition function is not differentiable. This accords with common sense, because when $t = 1$ the interaction between different configurations is zero, and so the weight of all states which are not constant is zero.

There is an analogue of Theorem 3 for the order parameter: roughly speaking, in the case of a positive model $\mathcal{P}$, the order parameter of a graph scheme is identically zero.

Theorem 4. Let $\mathcal{P}[t] = (A, i[t])$ be a family of positive models, and let $G = (F, J)$ be a graph scheme such that the transfer matrix of G with respect to $\mathcal{P}[t]$ has a spectral decomposition as in (2. 4. 4). Then the order parameter of G is zero for all $t > 0$.

Proof. As in Theorem 3, the proof leans heavily on the Frobenius property of positive matrices (Appendix B). Putting $\lambda_1 = \lambda_{max}$ in (2. 4. 5) and using the fact that $Z = \Sigma\lambda_i^n$, we obtain

$$\gamma_k(G) = \lim_{n\to\infty} \langle c_k \rangle_n = a_{11} + \sum_{j>2} a_{1j}(\lambda_j/\lambda_1)^k,$$

$$\gamma_\infty(G) = \lim_{k\to\infty} \gamma_k(G) = a_{11}.$$

We must show that $a_{11} = 0$, where, by definition

$$a_{11} = \sum_{(\sigma,\tau)} E_1(\sigma, \tau)E_1(\tau, \sigma)\, C(\sigma, \tau).$$

Arranging the terms in the sum according to the value of $C(\sigma, \tau)$, we have

$$a_{11} = (m - 1)S(0) - \sum_{b\neq 0} S(b),$$

where, for each b in A, $S(b)$ denotes the sum of those terms $E_1(\sigma, \tau)E_1(\tau, \sigma)$ for which $\sigma(x) - \tau(y) = b$. Thus it suffices to show that $S(b) = S(0)$ for each b.

Let u and v be left and right eigenvectors associated with the eigenvalue λ_1 of T; that is

$$u'T = \lambda_1 u', \qquad Tv = \lambda_1 v.$$

If we suppose that u and v are normalised so that $u'v = 1$, then, since λ_1 has unit multiplicity, u and v are uniquely determined and

$$E_1(\sigma, \tau) = u(\tau)v(\sigma).$$

Let $\sigma^{(b)}$ denote the state derived from σ by adding b to each value: $\sigma^{(b)}(w) = \sigma(w) + b$. From the definition of T it follows that

$$T(\sigma^{(b)}, \tau^{(b)}) = T(\sigma, \tau) \text{ for all } b \text{ in } A.$$

Define the permutation matrix P_b as follows: $P_b(\sigma, \tau) = 1$ if $\sigma = \tau^{(b)}$, and $P_b(\sigma, \tau) = 0$ otherwise. Then the preceding equation may be expressed by saying that T commutes with P_b, $P_bT = TP_b$. Hence,

$$T(P_bv) = P_bTv = P_b\lambda_1 v = \lambda_1(P_bv),$$

and so P_bv is an eigenvector of T, associated with λ_1. Since λ_1 has unit multiplicity, there is some constant k_b such that $P_b v = k_b v$, and since P_b is a permutation matrix of finite order, k_b is a root of unity. The Frobenius theorem guarantees that v is real and all its entries have the same sign, hence $k_b = 1$, and $P_b v = v$; in other words, for each state σ and each b in A,

$$v(\sigma^{(b)}) = v(\sigma) .$$

Clearly, the same is true for u.

Fix the state σ and let $\Sigma^{(b)}$ denote a sum over those states τ for which $\tau(y) = \sigma(x) - b$. Then $S(b) = \sum_{\sigma} S(b, \sigma)$, where

$$\begin{aligned} S(b, \sigma) &= \Sigma^{(b)} E_1(\sigma, \tau) E_1(\tau, \sigma) \\ &= \Sigma^{(b)} u(\tau)v(\sigma)u(\sigma)v(\tau) \\ &= u(\sigma)v(\sigma)\Sigma^{(b)} u(\tau)v(\tau) . \end{aligned}$$

But since $u(\tau)v(\tau)$ is the same as $u(\tau^{(b)})v(\tau^{(b)})$, the sum is the same as $\Sigma^{(0)}$. Hence

$$S(b) = \sum_{\sigma} S(b, \sigma) = \sum_{\sigma} S(0, \sigma) = S(0) ,$$

and $a_{11} = 0$, as required. //

The absence of transition points, for positive models on the graphs arising from a graph scheme G, is related to the one-dimensional nature of the sequence $\{G_n\}$. In two dimensions it is found that transition points do occur, even for positive models. We have already quoted the example of the Ising model on the plane square lattice, and we shall discuss it again in Chapter 5.

NOTES AND REFERENCES FOR CHAPTER 2

The expansion (2.1.10) of the partition function of the model $\mathcal{C}$ (in other words, the chromatic polynomial) was obtained by Whitney [9]; Birkhoff [2] had given a similar formula for map-colourings as long ago as 1912. In theoretical physics, expansions of this kind were found by Mayer [6] and others in the 1930s, while the high-temperature expansion of the Ising model is due to van der Waerden [8]. The fact that certain partition functions may be expanded in terms of the rank polynomial (2.1.9) was not noticed until fairly recently (see Essam [3]), and the notion of a resonant model is introduced here to make the relationship clear. In general, a graph function which is expressible as a rank polynomial will also have a deletion and contraction property, similar to (1.4.4). C. W. Vout [10] has shown that, in the case of the partition function of an interaction model, such properties are equivalent to a weak form of the resonance property.

The transfer matrix approach to the calculation of partition functions dates back to Kramers and Wannier [5]. They applied the method to the Ising model on the plane square lattice, and their arguments led to the determination of the transition point in that case. The method has since been widely used in physical problems, and even (occasionally) in graph-theoretical investigations [1].

The non-existence of transition points in one-dimensional physical systems, as a consequence of the Frobenius theorem, is well-understood (Ruelle [7, Section 5.6]). In fact, Kac [4] suggested that the degeneracy properties of eigenvalues provide a general mathematical explanation for the existence of transition points.

1. Biggs, N. L. Colouring square lattice graphs. Bull. London Math. Soc., 9 (1977), 54-6.
2. Birkhoff, G. D. A determinant formula for the number of ways of coloring a map. Ann. Math., 14 (1912), 42-6.
3. Essam, J. W. Graph theory and statistical physics. Discrete Math., 1 (1971), 83-112.
4. Kac, M. Mathematical mechanisms of phase transitions. Brandeis University Summer Institute in Theoretical Physics 1966, Volume 1, pp. 243-305. (Gordon and Breach, New York, 1968.)
5. Kramers, H. A. and Wannier, G. H. Statistics of the two-dimensional ferromagnet, I and II. Phys. Rev., 60 (1941), 252-76.
6. Mayer, J. E. Statistical mechanics of condensing systems. Part I. J. Chem. Phys., 5 (1937), 67-73.
7. Ruelle, D. Statistical Mechanics. (Benjamin, Reading (Mass.), 1969.)

8. van der Waerden, B. L. Die lange Reichweite der regelmassigen Atomanordnung in Mischkristallen. <u>Z. Physik.</u>, 118 (1941), 473.

9. Whitney, H. A logical expansion in mathematics. <u>Bull. Amer. Math. Soc.</u>, 38 (1932), 572-9.

10. Vout, C. W. Interaction models with the additive property. (In preparation.)

3 · Duality

3.1 Flows on a graph

There are many physical problems which may be formulated in terms of a 'flow' on a graph. In this chapter we shall examine some problems of this kind, and show how they may be related to interaction models.

We recall from Chapter 1 the definition of the boundary operator $\partial : \Phi(G, A) \to \Omega(G, A)$, as follows:

$$(\partial\phi)(v) = \sum_{e \in EG} D_{ve}\phi(e) ,$$

where D is the incidence matrix of G with respect to some fixed orientation. An element ϕ in $\Phi(G, A)$ such that $\partial\phi$ is the zero state - that is, for which

$$(\partial\phi)(v) = 0 \text{ for all } v \text{ in } VG,$$

will henceforth be referred to as a flow on G with values in A. (The more general use of the word 'flow' in Chapter 1 was a temporary expedient, and is now superseded.) Fig. 8 represents a flow with values in A_5, on the graph with orientation shown.

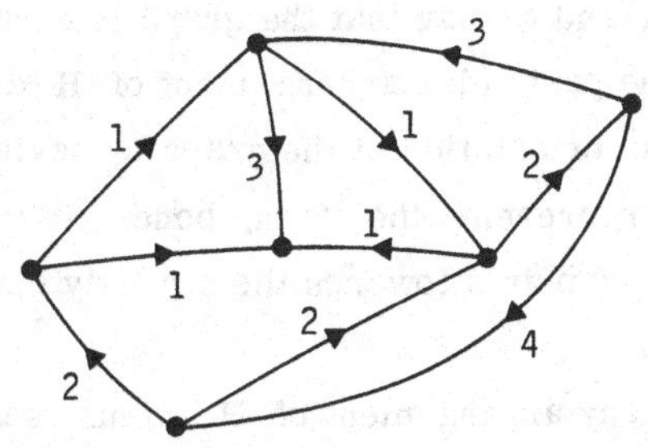

Fig. 8. A flow modulo 5

The set of all flows on G with values in A will be denoted by K or K(G, A). If ϕ_1 and ϕ_2 are in K, then so is $\phi_1 + \phi_2$.

We shall now digress to explain a famous problem which involves counting flows of a certain kind. A block of ice contains oxygen atoms and hydrogen atoms, each oxygen atom being strongly bonded to two hydrogens according to the molecular formula H_2O. The O atoms have some thermal motion about their equilibrium positions - in real ice these positions are the lattice points of a hexagonal wurtzite structure. As the temperature approaches absolute zero, the thermal oscillations die away, and one would expect that the 'residual entropy' of ice would be zero. However, measurements of specific heat indicate that this is not so. The explanation lies in the behaviour of the H atoms; in addition to its molecular bond with one O atom, each H atom has another, weaker, bond with some other O atom. We may think of the H atom as lying between the two O's, but closer to one of them. The H atoms may switch their positions, interchanging the strong and weak bonds, subject only to the so-called 'ice condition': each O has two strong bonds and two weak bonds. This uncertainty accounts for the residual entropy of ice.

The problem of counting the number of ways in which the hydrogens may be arranged can be expressed in terms of flows. We define a graph whose vertices correspond to the O atoms and whose edges correspond to the H atoms. The edge representing a particular H atom joins the vertices representing the two O atoms with which it is bonded, one bond being strong and the other one weak. It is clear that each vertex of this graph has valency four, and we say that the graph is 4-valent.

Let us select one particular arrangement of H atoms (call it the base level) and define an orientation of the graph by saying that the positive end of an edge represents the strong bond. In the diagrammatic representation, the arrow points towards the strongly bonded oxygen atom (see Fig. 9).

Now let us take any arrangement of H atoms, satisfying the ice condition, and assign the value $\phi(e) = +1$ to the edge e if the corresponding H atom is in the same position as in the base level, but $\phi(e) = -1$

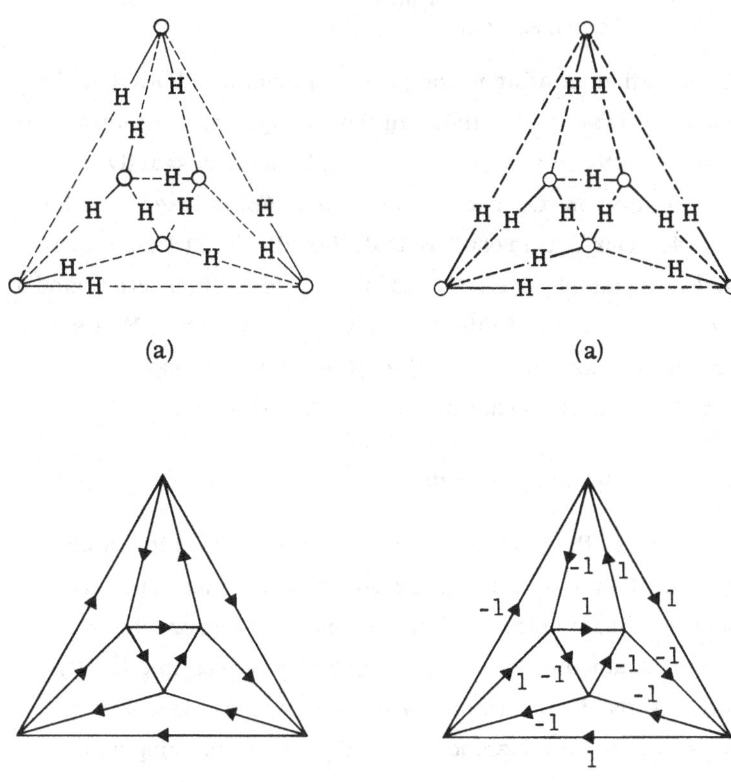

(b) (b)

Fig. 9. The base level and the oriented graph

Fig. 10. Another arrangement and the corresponding flow

if it has been switched (see Fig. 10). We shall treat the numbers +1 and -1 as residues modulo 3, when it is clear that ϕ is a flow with values in A_3. We refer to this kind of flow as a **proper flow**, since it never takes the value zero.

In summary, we have formulated the problem of the residual entropy of ice in the following way: take a 4-valent graph G, and count the number of proper flows on G with values in A_3.

The ice problem may be formulated within the following general framework. Let G be a graph and $\mathfrak{M}$ an interaction model; define

$$Y(\mathfrak{M}, G) = \sum_{\phi \in K(G, A)} \prod_{e \in E} i[\phi(e)] . \tag{3.1.1}$$

Although Y is rather similar to the partition function $Z(\mathfrak{M}, G)$, the definitions are not exactly parallel. In the first place, the interaction i is evaluated directly in terms of ϕ, which is not necessarily the coboundary of a state on G; and secondly, the sum is taken over the subset K of Φ. Our main result is that, for all G, $Y(\mathfrak{M}, G)$ can be expressed in terms of $Z(\hat{\mathfrak{M}}, G)$, where $\hat{\mathfrak{M}}$ is an interaction model derived from $\mathfrak{M}$. In view of this result, we shall not give Y a special name, but refer to it as the 'Y function' wherever necessary.

When $\mathfrak{M}$ is the resonant model $\mathcal{C} = (A, i)$ with

$$i(0) = 0, \quad i(a) = 1 \qquad (a \neq 0) ,$$

the Y function gives the number of proper flows on G with values in A. In particular, if G is a 4-valent graph and $A = A_3$, then $Y(\mathcal{C}, G)$ is a measure of the residual entropy of the piece of ice represented by G.

Another instance of the function $Y(\mathcal{C}, G)$ arises in a totally different way. A graph G in which each vertex is incident with just three edges is said to be **trivalent.** Let F_4 denote the ring with four elements $0, 1, \lambda, \mu$, and operations defined by the tables:

+	0	1	λ	μ
0	0	1	λ	μ
1	1	0	μ	λ
λ	λ	μ	0	1
μ	μ	λ	1	0

·	0	1	λ	μ
0	0	0	0	0
1	0	1	λ	μ
λ	0	λ	μ	1
μ	0	μ	1	λ

From the addition table we note that $x + x = 0$, or $x = -x$, for each x in F_4, so that F_4 is not the same as the ring A_4 (in fact, F_4 is a field). Suppose that ϕ is a proper flow on the trivalent graph G, with values in F_4, and let e_1, e_2, e_3, be the three edges incident with some given vertex v of G. The condition $\partial\phi(v) = 0$ means that

$$\phi(e_1) + \phi(e_2) + \phi(e_3) = 0, \tag{3.1.2}$$

taking account of the fact that + and - are the same in F_4. For a proper flow, each $\phi(e_i)$ is non-zero, and it is easy to check that the only way that (3.1.2) can be satisfied is for the three summands to be 1, λ, and μ in some order. In other words, a proper flow assigns one of the three 'colours' 1, λ, μ, to each edge of G, in such a way that all three colours occur at each vertex. This is known as an 'edge-3-colouring' or 'Tait colouring' of G, and so we may say that, in this case, $Y(\mathcal{C}, G)$ is equal to the number of Tait colourings of G. In the case of a trivalent planar graph (Section 3.4), a Tait colouring is equivalent to a proper-4-colouring of the <u>vertices</u> of the dual graph. Consequently, the four-colour problem may be expressed in terms of the Y function.

3.2 Dual models

In this section we shall introduce an algebraic theory of duality for interaction models. It must be stressed at the outset that this theory is quite independent of the well-known concept of duality for planar graphs, which we shall consider in a later section.

Let U denote the set of complex numbers of unit modulus. The usual multiplication of complex numbers is a group operation on U, and it is commutative, so that U is an abelian group, with identity element 1.

The ring A which appears in an interaction model $\mathfrak{M} = (A, i)$ is also an abelian group, if we restrict our attention to the addition operation in A. A character of A is a function $\chi : A \to U$ which has the property

$$\chi(a + b) = \chi(a)\chi(b) \text{ for all } a, b \text{ in } A.$$

In other words, χ is a homomorphism of the group A into U. For any character we must have $\chi(0) = 1$, and the character defined by $\chi(a) = 1$, for all a in A, is said to be **trivial**.

The characters of A_m are easily determined. For each integer $l = 0, 1, \ldots, m-1$ we have a character χ_l defined by

$$\chi_l(b) = \exp(2\pi i b l/m),$$

where the symbol b on the right-hand-side stands for any integer belonging to the residue class b in A_m. We note that χ_0 is the trivial character.

In the ensuing theory we shall need the ring structure of A, and we shall use characters which have a simple property with respect to this structure. The character χ is said to be **ring-like** if it is non-trivial and

$$\sum_{b \in A} \chi(ab) = 0 \qquad (a \neq 0). \tag{3.2.1}$$

When $A = A_m$ and l, m are coprime, the character χ_l is ring-like, because of elementary facts about roots of unity. In the general case it is convenient to postulate (3.2.1) explicitly.

Let $\mathfrak{M} = (A, i)$ be an interaction model, and χ any ring-like character on A. The **dual** interaction model $\hat{\mathfrak{M}}$ is the model $(\hat{A}, \hat{i})$, where $\hat{A} = A$ and $\hat{i}$ is defined by

$$\hat{i}(a) = |A|^{-\frac{1}{2}} \sum_{b \in A} i(b)\chi(-ab) . \tag{3.2.2}$$

(A more sophisticated definition is possible, in which $\hat{A}$ is not the same as A, but merely isomorphic with it.) We remark that $\hat{i}$ is a kind of 'finite Fourier transform' of i.

In the case of a resonant model, the dual model is also resonant. To see this, suppose that $\mathfrak{R} = (A, i)$, where i has the general resonant form as given in (2.1.1). Then

$$\begin{aligned} \hat{i}(a) &= |A|^{-\frac{1}{2}} \sum_{b \in A} i(b)\chi(-ab) \\ &= |A|^{-\frac{1}{2}} \{i_0\chi(0) + i_1 \sum_{b \neq 0} \chi(-ab)\} \\ &= |A|^{-\frac{1}{2}} \{i_0 + i_1 \sum_{b \in A} \chi(-ab) - i_1\}. \end{aligned}$$

If $a = 0$, the sum Σ is equal to $|A|$, whereas if $a \neq 0$ it is zero, by the ring-like property. Hence $\hat{i}$ has the resonant form, with

$$\hat{i}_0 = \frac{i_0 + (m-1)i_1}{\sqrt{m}} , \qquad \hat{i}_1 = \frac{i_0 - i_1}{\sqrt{m}} , \tag{3.2.3}$$

where we have used our standard abbreviation $m = |A|$. The equations (3.2.3) show that, in the case of a resonant model $\mathcal{R}$, the dual $\hat{\mathcal{R}}$ is unique and does not depend on the character χ. In the case of a general model $\mathfrak{M}$ it should be remembered that $\hat{\mathfrak{M}}$ may depend on the choice of χ, even though we do not mention χ explicitly. However, the double dual $\hat{\hat{\mathfrak{M}}}$ is independent of χ, since it is in fact the same as $\mathfrak{M}$. To see this, we compute $\hat{\hat{i}}$, as follows:

$$\hat{\hat{i}}(a) = |A|^{-\frac{1}{2}} \sum_{b\epsilon A} \hat{i}(b)\chi(-ab)$$

$$= |A|^{-\frac{1}{2}} \sum_{b\epsilon A} |A|^{-\frac{1}{2}} \sum_{c\epsilon A} i(c)\chi(-bc)\chi(-ab) .$$

Since χ is a character and multiplication in A is commutative, we have $\chi(-bc)\chi(-ab) = \chi(b(-c-a))$. Furthermore, since χ is ring-like,

$$\sum_{b\epsilon A} \chi(b(-c - a)) = \begin{cases} |A| & \text{if } c = -a, \\ 0 & \text{otherwise.} \end{cases}$$

Hence, rearranging the sum above, we obtain just one term

$$\hat{\hat{i}}(a) = i(-a) .$$

But $i(-a) = i(a)$ by the definition of an interaction model, and so $\hat{\hat{\mathfrak{M}}} = \mathfrak{M}$.

It is worth noting that the transform (3.2.1) may be inverted by the formula

$$i(a) = |A|^{-\frac{1}{2}} \sum_{b\epsilon A} \hat{i}(b)\chi(ab).$$

This is a simple consequence of the ring-like property.

3.3 The algebraic duality theorem

We prepare for the main result of this chapter (Theorem 5) by proving some simple identities involving the boundary and coboundary operators.

Given any ω in $\Omega(G, A)$ and any ϕ in $\Phi(G, A)$ we have an element $\omega . \partial\phi$ of Ω and an element $\delta\omega . \phi$ of Φ, related by the identity

$$\sum_{e\in E} (\delta\omega . \phi)(e) = \sum_{v\in V} (\omega . \partial\phi)(v) . \qquad (3.3.1)$$

To prove this, we consider the double sum

$$\sum_{e\in E} \sum_{v\in V} D_{ve}\omega(v)\phi(e) .$$

Evaluating the sums in the order shown yields the left-hand-side of (3.3.1), while interchanging the order yields the right-hand-side.

Let χ be any ring-like character of A. Applying χ to both sides of (3.3.1) and using the homomorphism property, we obtain

$$\prod_{e\in E} \chi[(\delta\omega . \phi)(e)] = \prod_{v\in V} \chi[(\omega . \partial\phi)(v)] . \qquad (3.3.2)$$

Summing (3.3.2) over all states ω,

$$\sum_{\omega\in\Omega} \prod_{e\in E} \chi[(\delta\omega . \phi)(e)] = \sum_{\omega\in\Omega} \prod_{v\in V} \chi[(\omega . \partial\phi)(v)] .$$

Now we can use our useful Lemma I (Appendix A). Since Ω is just the set of functions $\omega : V \to A$, the second expression above is equal to

$$\prod_{v\in V} \sum_{a\in A} \chi[a . (\partial\phi)(v)].$$

The ring-like property of χ means that the sum is zero, unless $(\partial\,\phi)(v) = 0$, when it takes the value $|A|$. Hence the whole expression is zero, unless $\partial\phi = 0$, when it takes the value $|A|^{|V|}$. Consequently,

$$\sum_{\omega\in\Omega} \prod_{e\in E} \chi[\partial\omega(e) . \phi(e)] = \begin{cases} |A|^{|V|} & \text{if } \partial\phi = 0; \\ 0 & \text{otherwise.} \end{cases} \qquad (3.3.3)$$

Theorem 5. <u>Suppose that $\hat{\mathfrak{M}}$ is the dual of an interaction model $\mathfrak{M} = (A, i)$ with respect to a character χ of A, and let $Y(\mathfrak{M}, G)$ be as defined in (3.1.1). Then for any graph G,</u>

$$Y(\mathfrak{M}, G) = m^{\frac{1}{2}|E|} \overline{Z}(\hat{\mathfrak{M}}, G), \qquad (3.3.4)$$

<u>where</u> $m = |A|$.

Proof. From the definitions of Z and $\hat{\mathfrak{M}}$, we have

$$Z(\hat{\mathfrak{M}}, G) = \sum_{\omega} \prod_{e} \hat{i}[\delta\omega(e)]$$
$$= \sum_{\omega} \prod_{e} m^{-\frac{1}{2}} \sum_{b} i(b)\chi[-\delta\omega(e). b]$$
$$= m^{-\frac{1}{2}|E|} \sum_{\omega} \prod_{e} \sum_{b} i(b)\chi[-\delta\omega(e). b].$$

Rearranging, and applying Lemma I to the $\Pi\Sigma$ term,

$$m^{\frac{1}{2}|E|} Z(\hat{\mathfrak{M}}, G) = \sum_{\omega} \sum_{\phi\epsilon\Phi} \prod_{e} i[\phi(e)]\chi[-\delta\omega(e)\phi(e)]$$
$$= \sum_{\phi} \sum_{\omega} \prod_{e} i[\phi(e)]\chi[-\delta\omega(e)\phi(e)].$$

Now the Πi product does not depend on ω, and so it may be moved to the left of the sum over ω, giving

$$m^{\frac{1}{2}|E|} Z(\hat{\mathfrak{M}}, G) = \sum_{\phi} \prod_{e} i[\phi(e)] \sum_{\omega} \prod_{e} \chi[-\delta\omega(e)\phi(e)].$$

By (3. 3. 3), the final $\Sigma\Pi$ term is zero unless $\partial\phi = 0$, when it takes the value $m^{|V|}$. Hence

$$m^{\frac{1}{2}|E|} Z(\hat{\mathfrak{M}}, G) = m^{|V|} \sum_{\phi\epsilon K} \prod_{e} i[\phi(e)] = m^{|V|} Y(\mathfrak{M}, G).$$

Dividing through by $m^{|V|}$ we obtain the stated result. //

Theorem 5 may be viewed in two ways. As stated, it enables us to reduce questions like the ice problem, or the Tait colouring problem, to the standard form of evaluating the partition function of an interaction model. The dual of the model $\mathfrak{C}$ is $\hat{\mathfrak{C}} = (A, \hat{i})$, where $\hat{i}$ is obtained from (3. 2. 3) with $i_0 = 0$ and $i_1 = 1$; that is,

$$\hat{i}_0 = \sqrt{m} - \frac{1}{\sqrt{m}}, \qquad \hat{i}_1 = -\frac{1}{\sqrt{m}}.$$

In order to clear away the square roots, let $\mathfrak{C}'$ be the resonant model (A, i') with

$$i'_0 = m - 1, \qquad i'_1 = -1.$$

Then $\overline{Z}(\mathcal{C}', G) = m^{\frac{1}{2}|E|}\overline{Z}(\hat{\mathcal{C}}, G)$ and so the number of proper flows on G is given by

$$Y(\mathcal{C}, G) = \overline{Z}(\mathcal{C}', G) . \qquad (3.3.5)$$

General properties of the function Y, such as the existence of subgraph expansions similar to those discussed in Section 2.1, may be inferred immediately from Theorem 5. In particular, applying the rank polynomial expansion (2.1.9) to the right-hand-side of (3.3.5), we obtain

$$\begin{aligned} Y(\mathcal{C}, G) &= (-1)^{|E|} \sum_{S \subseteq E} (-m)^{|S|} \left(\frac{1}{m}\right)^{r\langle S\rangle} \\ &= (-1)^{|E|} \sum_{S \subseteq E} (-1)^{|S|} m^{r^*\langle S\rangle} , \end{aligned} \qquad (3.3.6)$$

where

$$r^*(H) = |EH| - r(H) = |EH| - |VH| + c(H)$$

is known as the co-rank, or cyclomatic number of the graph H. (It is the number of independent circuits in H.) Incidentally, the relation (3.3.6) shows that $Y(\mathcal{C}, G)$, like $Z(\mathcal{C}, G)$, is a polynomial function of m.

Another view of Theorem 5 is that it provides an alternative approach to the evaluation of a partition function. For example, when A is the ring A_2 the Y function has an especially simple form. Each ϕ in $\Phi(G, A_2)$ can take only the values 0 and 1, and so there is a one-to-one correspondence $\phi \leftrightarrow S_\phi$ between Φ and the set of subsets S of EG, given by

$$S_\phi = \{e \in EG \mid \phi(e) = 1\}.$$

So for any model $\mathcal{R} = (A_2, i)$, necessarily resonant,

$$\prod_{e \in E} i[\phi(e)] = i_0^{|E-S_\phi|}\, i_1^{|S_\phi|} = i_0^{|E|}(i_1/i_0)^{|S_\phi|} .$$

The flows (elements of $K(G, A_2)$) correspond to those subsets S for which $\langle S\rangle$ has an even number of edges at each vertex - the admissible edge-subgraphs, in the terminology of Section 2.1. Thus Theorem 5 leads to the formula

$$\overline{Z}(\mathcal{R}, G) = 2^{-\frac{1}{2}|E|} Y(\hat{\mathcal{R}}, G)$$

$$= 2^{-\frac{1}{2}|E|} \hat{i}_0^{|E|} \sum_{\substack{S \\ \text{admissible}}} (\hat{i}_1/\hat{i}_0)^{|S|}$$

$$= \left(\frac{i_0 + i_1}{2}\right)^{|E|} \sum_{\substack{S \\ \text{admissible}}} \left(\frac{i_0 - i_1}{i_0 + i_1}\right)^{|S|} .$$

In particular, when $\mathcal{R}$ is the Ising model $\mathcal{I}$, with $\varepsilon = \exp \nu$, we recover the high-temperature expansion obtained by a different method in Section 2.1.

3.4 Planarity and duality

In this section we shall consider the implications of a rather special kind of duality: it is geometric, rather than algebraic, and it applies to graphs, not interactions.

When we represent a graph by a diagram of points and lines (as in Fig. 1) we often find it necessary to make the lines cross at points which do not represent vertices (as, for example, in Fig. 5). We say that an abstract graph is **planar** if it can be represented by a plane diagram without extraneous crossings. For example, the complete graph K_4 is planar - the crossing point in Fig. 11(a) is not essential, and it can be removed by redrawing the graph as in (b). However, the complete graph K_5 is nonplanar, since it is impossible to draw it without at least

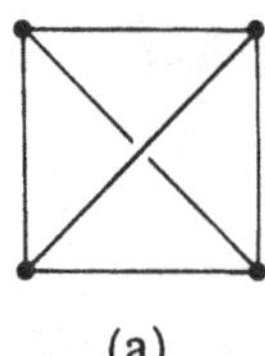

(a)

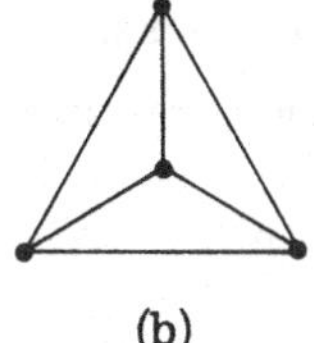

(b)

Fig. 11. K_4 is planar

one crossing. In our discussions of planarity and planar graphs we shall adopt a somewhat unsophisticated attitude to topological details, on the grounds that such details tend to obscure the important features.

Suppose that G is a connected planar graph. We may construct a dual graph G^* as follows: the vertices of G^* are in one-to-one correspondence with the regions into which the plane is divided by a planar representation of G, and the edges of G^* join vertices corresponding to adjacent regions, in such a way that there is a one-to-one correspondence between the edges e of G and the edges e^* of G^*. In a pictorial representation we place the vertices of G^* inside the corresponding regions of G, and draw the edge e^* so that it crosses the corresponding edge e at right angles. See Fig. 12.

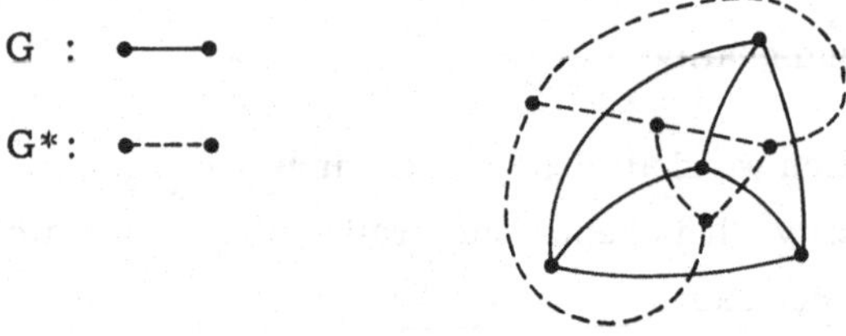

Fig. 12. Dual graphs

When G has an orientation, there is a compatible orientation of G^* obtained by applying the rule that the positive direction on e may be brought into coincidence with the positive direction on e^* by a clockwise rotation through $\frac{1}{2}\pi$. We shall always assume that G and G^* have compatible orientations, and denote their respective incidence matrices by D and D^*. The columns of D^* (edges of G^*) will be labelled in such a way as to correspond with the columns of D (edges of G). With these conventions, we may prove the important identity

$$DD^{*\prime} = 0. \qquad (3.4.1)$$

For the proof, consider the entry in the product matrix corresponding to a vertex v of G and vertex w^* of G^*; it is

$$\sum_e D_{ve} D^*_{w^*e^*} \cdot$$

A typical summand is zero unless v is incident with e and w* is incident with the dual edge e*, that is, unless v lies on the boundary of the region corresponding to w*. If v and w* are so related, then there are just two edges for which the summand is non-zero, and the rule for compatible orientations implies that one summand is +1 and the other is -1 (Fig. 13). Hence the sum is always zero.

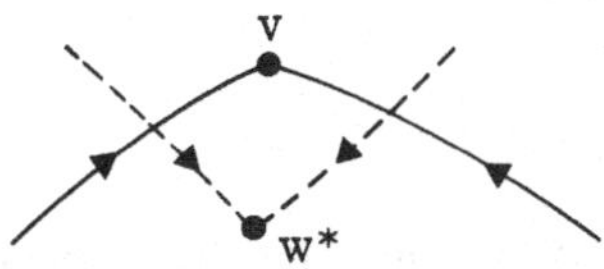

Fig. 13. Compatible orientations

The identity (3.4.1) may be translated into a result about the boundary and coboundary operators. The one-to-one correspondence between the edges e and e* leads to a one-to-one correspondence $f \longleftrightarrow f^*$ between the rings $\Phi(G, A)$ and $\Phi(G^*, A)$, defined by $f^*(e^*) = f(e)$. It will be convenient to identify f and f*, so that we have the following diagram of operators:

$$\begin{array}{ccc} \Phi(G, A) & \underset{\delta}{\overset{\partial}{\rightleftarrows}} & \Omega(G, A) \\ \| & & \\ \Phi(G^*, A) & \underset{\delta^*}{\overset{\partial^*}{\rightleftarrows}} & \Omega(G^*, A) \, . \end{array}$$

From the definitions of ∂ and δ (Section 1.2), and the equation (3.4.1), we deduce that if $f^* = \delta^*\omega^*$ then $\partial f = 0$. We shall require the converse of this result, which is a bit more subtle. (In fact, it is precisely the topological characterization of the plane or sphere.) For convenience we shall assume that the ring A is actually a field; this implies that the cardinality m must be a prime or a power of a prime. The point of the assumption is that Φ and Ω become vector spaces, while ∂ and δ

become linear transformations, and so familiar arguments concerning the dimensions of subspaces can be empolyed. We remark that ∂ is represented by the matrix D, and δ is represented by the transposed matrix D'; similarly for ∂^* and δ^*.

First, we determine the rank of D. Let d_v denote the row of D corresponding to the vertex v of G. Since each column of D contains just two non-zero entries, one +1 and one -1, we have the linear relation

$$\sum_v d_v = 0 . \qquad (3.4.2)$$

Consequently, the rows of D are not linearly independent, and the rank of D is at most $|V| - 1$; we shall show that it is exactly $|V| - 1$. Suppose that we have any linear relation

$$\sum_v \alpha_v d_v = 0 ,$$

where not all the coefficients α_v are zero. Choose a row d_u for which $\alpha_u \neq 0$; this row has non-zero entries in those columns corresponding to the edges incident with u. For each such column, there is just one other row d_w with a non-zero entry in that column, and, since the relevant entries are +1 and -1, if the given linear relation is to hold we must have $\alpha_u = \alpha_w$. Consequently, $\alpha_u = \alpha_v$ for all vertices v adjacent to u. The same argument applies to all vertices adjacent to vertices which themselves are adjacent to u, and so on. Since G is assumed to be connected, all vertices can be covered in this way, and $\alpha_u = \alpha_v$ for all vertices of G. In other words, the given linear relation is just a multiple of (3.4.2), and D has $|V| - 1$ linearly independent rows.

It is clear that G* is connected whenever G is, so that the rank of D* is $|V^*| - 1$. At this point we appeal to the classical formula of Euler concerning the numbers of vertices, edges, and regions of a plane network:

$$|V| - |E| + |V^*| = 2, \qquad (3.4.3)$$

where $|V^*|$, the number of vertices of G^*, is equal to the number of regions in a plane representation of G. Hence

$$\text{rank } D + \text{rank } D^* = |E|. \tag{3.4.4}$$

We now reconsider the relationship between the space

$$I^* = \{f^* \in \Phi(G^*, A) \,|\, f^* = \delta^*\omega^* \text{ for some } \omega^* \text{ in } \Omega(G^*, A)\},$$

and the space $K(G, A)$ of flows on G. We have already seen that, with identification of f and f^*, I^* is a subspace of K. Now the dimension of I^* is equal to the rank of the matrix $D^{*\prime}$ which represents δ^*, and this is the same as the rank of D^*. The dimension of K is given by a well-known theorem in elementary linear algebra:

$$\dim K + \text{rank } D = \dim \Phi.$$

Since $\dim \Phi = |E|$, it follows from (3.4.4) that

$$\dim K = |E| - \text{rank } D = \text{rank } D^*.$$

Consequently, since I^* and K have the same dimension, we must have $I^* = K$. In other words, if $\partial f = 0$, then $f^* = \delta^*\omega^*$ for some ω^*.

Theorem 6. Let G be a connected planar graph and A a field with m elements. For any model $\mathfrak{M} = (A, i)$, the partition function of $\mathfrak{M}$ on the dual graph G^* is given by

$$Z(\mathfrak{M}, G^*) = m\, Y(\mathfrak{M}, G). \tag{3.4.5}$$

Proof. We shall require a simple, but precise, specification of the kernel of δ^*: that is, the space of states σ on G^* for which $\delta^*\sigma = 0$. Each of the constant states τ_a, taking the value a at every vertex of G^*, belongs to the kernel; there are m such states. Since δ^* is a transformation of rank $|V^*| - 1$ on a space of dimension $|V^*|$, its kernel has dimension 1, and so it consists precisely of these m states.

We have seen that if f is a flow, then $f^* = \delta^*\omega^*$ for some ω^*; in fact, there are just m such states ω^*. For suppose ω^*_0 is one of them, and σ is another. Then

$$\delta^*(\omega^*_0 - \sigma) = \delta^*\omega^*_0 - \delta^*\sigma = f^* - f^* = 0.$$

Hence $\omega^*_0 - \sigma$ belongs to the kernel of δ^*, and

$$\omega^*_0 - \sigma = \tau_a \quad \text{for some } a \text{ in } A.$$

Writing this in the form $\sigma = \omega^*_0 + \tau_a$ we conclude that there are m states σ, as claimed.

The preceding arguments show that the transformation δ^* (together with the identification of f and f*) establishes an m-to-one correspondence between the states on G* and the flows on G. By definition,

$$Z(\mathfrak{M}, G^*) = \sum_{\omega^*} \prod_{e^*} i[\delta^*\omega^*(e^*)].$$

Because of the m-to-one correspondence, the right-hand-side is equal to

$$m \sum_{f \in K} \prod_{e} i[f(e)] ,$$

which is just $mY(\mathfrak{M}, G)$. //

A simple example of (3. 4. 5) is provided by taking $\mathfrak{M}$ to be the colouring model $\mathfrak{C}$. We deduce that the number of proper colourings of G*, with m colours available, is equal to m times the number of proper flows on G. In particular, the 'residual entropy' associated with a planar 4-valent graph may be expressed in terms of the number of proper 3-colourings of the dual graph.

We now have two duality statements, (3. 3. 4) and (3. 4. 5), and it is natural to put the two together. When the hypotheses of Theorems 5 and 6 hold, we have

$$Z(\hat{\mathfrak{M}}, G) = m^{|V| - \frac{1}{2}|E| - 1} Z(\mathfrak{M}, G^*). \qquad (3.4.6)$$

Euler's formula (3. 4. 3) implies that

$$\{|V| - \tfrac{1}{2}|E| - 1\} + \{|V^*| - \tfrac{1}{2}|E^*| - 1\} = 0,$$

so that the equation (3. 4. 6) has perfect symmetry with respect to G and G*. This equation has been employed to some effect in theoretical physics, as we shall explain in the next section. However, it is clear that it is really a consequence of the two more fundamental dualities, one algebraic and one geometric. This would tend to imply that the planarity of a graph is of less significance in physical problems than is sometimes thought. Another pointer to the same conclusion will occur in our discussion of dimensionality in Chapter 5.

3. 5 Transition points for planar graphs

When $\mathcal{R}$ is a resonant model, with interactions i_0 and i_1, the corresponding partition function is determined, up to a multiplicative factor, by the ratio $\rho = i_0/i_1$. This is an immediate consequence of the definition and is explicit, for example, in equation (2. 1. 9). We shall use the temporary notation

$$Z(\mathcal{R}, G) = i_1^{|E|}\, z(\rho, G),$$

where z depends on G and the single variable ρ. Similarly, the dual model $\hat{\mathcal{R}}$ is also a resonant model, and

$$Z(\hat{\mathcal{R}}, G) = i_1^{|E|}\, z(\hat{\rho}, G),$$

where $\hat{\rho} = \hat{i}_0/\hat{i}_1$ and the second z function is the same as the first, for the same graph G.

Now suppose that $\mathcal{R}[t]$ is a family of resonant models, and $\hat{\mathcal{R}}[t]$ is the dual of $\mathcal{R}[t]$. It may happen that for each $t > 0$ there is some $t^* > 0$ satisfying

$$\rho[t] = \hat{\rho}[t^*]. \qquad (3.5.1)$$

Using the explicit formulae (3. 2. 3) for $\hat{i}_0$ and $\hat{i}_1$, this is equivalent to

$$\frac{i_0[t]}{i_1[t]} = \frac{i_0[t^*] + (m-1)i_1[t^*]}{i_0[t^*] - i_1[t^*]}$$

which reduces to the symmetrical form

$$\{\rho[t] - 1\}\{\rho[t^*] - 1\} = m. \tag{3.5.2}$$

When t^* exists, we may use (3.5.1) to relate $\mathcal{R}[t]$ and $\hat{\mathcal{R}}[t^*]$ on G. Since $z(\rho[t], G) = z(\hat{\rho}[t^*], G)$, we have

$$Z(\mathcal{R}[t], G) = x[t]^{|E|} Z(\hat{\mathcal{R}}[t^*], G),$$

where $x[t] = i_1[t]/i_1[t^*]$. When, in addition, G is planar, we may use (3.4.6) to express the right-hand-side in terms of the partition function of $\mathcal{R}[t^*]$ on the dual graph G^*. We obtain

$$Z(\mathcal{R}[t], G) = x[t]^{|E|} m^{|V| - \frac{1}{2}|E| - 1} Z(\mathcal{R}[t^*], G^*). \tag{3.5.3}$$

This equation, although it is based on several assumptions, has at least one interesting application. The infinite plane square lattice graph ⊞ is self-dual (Fig. 14). We may choose a sequence of finite subgraphs of ⊞ (for example, take G_n to be an $n \times n$ square subgraph) so that the dual sequence $\{G_n^*\}$ also approaches ⊞ . So we may write, for the partition function of ⊞ ,

$$\begin{aligned} Z_\infty[t] &= \lim_{n\to\infty} \{Z(\mathcal{R}[t], G_n)\}^{1/v_n} \\ &= \lim_{n\to\infty} \{Z(\mathcal{R}[t], G_n^*)\}^{1/v_n^*}. \end{aligned}$$

Now $v_n/v_n^* \to 1$ as $n \to \infty$; also, if e_n denotes the number of edges of G_n, then $e_n/v_n \to 2$ as $n \to \infty$, since ⊞ has valency four. Hence, taking limits in (3.5.3), we obtain

$$Z_\infty[t] = x[t]^2 Z_\infty[t^*]. \tag{3.5.4}$$

It follows from (3.5.4) that transition points on ⊞ occur in pairs; if t_0 is a transition point, then so is t_0^*. In particular, if there is a unique transition point it must occur when $t = t^*$, and so it is determined by (3.5.2):

$$\rho[t_c] = 1 + \sqrt{m}.$$

We remark that the 'normalization factor' $x[t]$ is equal to 1 when $t = t^*$; this is a consequence of (3.5.2) and the definition of $x[t]$, and is confirmed by equation (3.5.4).

The significance of the preceding arguments lies in the fact that the Ising model exhibits precisely the postulated behaviour, so that its transition point on ⊞ may be located. There is no extra work involved in dealing with the family of Potts models: these are the resonant models with $m \geqq 2$ and

$$i_0[T] = \exp(L/kT), \quad i_1[T] = \exp(-L/kT).$$

The Ising model is the case $m = 2$. The equation relating T and T^* is

$$\{\exp(2L/kT) - 1\}\{\exp(2L/kT^*) - 1\} = m, \qquad (3.5.5)$$

and the transition point T_c is given by

$$\exp(2L/kT_c) = 1 + \sqrt{m}\,.$$

These arguments are very special to the graph ⊞ , which is an infinite self-dual planar graph. However, the technique can be adapted to yield a relationship between transition points (if they exist) on dual pairs of infinite graphs. For instance, the critical temperature T_1 on the triangular lattice is related to the critical temperature T_2 on the

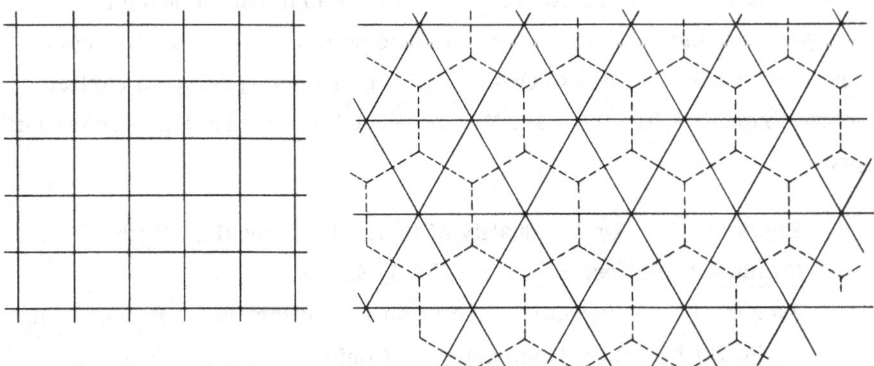

Fig. 14. (a) The square lattice (self-dual) (b) The triangular lattice and its dual - the hexagonal lattice

dual hexagonal lattice (Fig. 14) by

$$\{\exp(2L/kT_1) - 1\}\{\exp(2L/kT_2) - 1\} = m .$$

NOTES AND REFERENCES FOR CHAPTER 3

The accepted explanation for the residual entropy of ice originated in the 1930s; see Pauling [4]. The exact calculation of the thermodynamic limit in 'square ice' (that is, for the infinite plane square lattice) was carried through by Lieb [3], who also remarked on the equivalence, attributed to Lenard, with the three-colouring problem.

The result on algebraic duality (Theorem 5) is quite recent [1]. Tutte [5] studied the properties of flows with values in an abelian group, and formulated the special case of Theorem 6 for the model $\mathcal{C}$. He also conjectured [6] that every graph with a simple connectivity property has a proper flow over A_5. This is a deep conjecture and so far the only progress is the replacement of A_5 by A_8 (Jaeger [2]). The contrast with the situation for proper colourings is worth noting: there is no such absolute bound for the number of colours required in a proper colouring.

Whitney [8] investigated the rank and co-rank of dual planar graphs, and transformed the results into an abstract definition of duality. His methods yield an alternative proof of (3.4.6) for resonant models, via the rank polynomial expansion (2.1.9). The determination of the transition point for the Ising model on the plane square lattice, by means of a duality argument, is due to Onsager [7]; he re-interpreted an earlier algebraic result of Kramers and Wannier which used their transfer matrix approach.

1. Biggs, N. L. On the duality of interaction models. Math. Proc. Cambridge Philos. Soc., 80 (1976), 429-36.
2. Jaeger, F. On nowhere-zero flows in multigraphs. Proceedings of the Fifth British Combinatorial Conference, pp. 373-8. (Utilitas Mathematica, Winnipeg, 1976.)
3. Lieb, E. H. Residual entropy of square ice. Phys. Rev., 162 (1967), 162-71.

4. Pauling, L. The structure and entropy of ice and of other crystals with some randomness of atomic arrangement. J. Amer. Chem. Soc., 57 (1935), 2680-4.

5. Tutte, W. T. A ring in graph theory. Proc. Cambridge Philos. Soc., 43 (1947), 26-40.

6. Tutte, W. T. On the imbedding of linear graphs in surfaces. Proc. London Math. Soc., 51 (1949), 474-83.

7. Wannier, G. H. The statistical problem in cooperative phenomena. Rev. Mod. Phys., 17 (1945), 50-60.

8. Whitney, H. Non-separable and planar graphs. Trans. Amer. Math. Soc., 34 (1932), 339-63.

4 · Expansions

4.1 Graph types

In Section 2.1 we obtained some formulae for the reduced partition function $\overline{Z}(\mathcal{R}, G)$ of a resonant model in terms of subgraphs of G. It is convenient to simplify matters by taking a 'normalized' model with $i_1 = 1$, so that, for example, the rank polynomial expansion (2.1.9) becomes

$$\overline{Z}(\mathcal{R}, G) = \sum_{S \subseteq EG} (i_0 - 1)^{|S|} (1/m)^{r\langle S \rangle}.$$

We may interpret such formulae in the following way: each set of interactions (or subset S of EG) contributes some quantity to the reduced partition function, and the contribution depends only on the structure of the edge-subgraph $\langle S \rangle$, not on the specific set S. For instance, in the high-temperature expansion of the Ising model, the contribution of any edge-subgraph with a vertex of odd valency is zero.

We propose to define the notion of a 'type', in such a way that graphs of the same type t make the same contribution $\theta(t)$. Suppose that this has been done. If we then rearrange a sum like the one displayed above, so that edge-subgraphs of the same type are bracketed together, we obtain a formula

$$\overline{Z}(\mathcal{R}, G) = \sum_{t} N(G, t)\, \theta(t),$$

where $N(G, t)$ is the number of subsets S of EG such that $\langle S \rangle$ is of type t, and $\theta(t)$ is the contribution from any subgraph of type t.

It is apparent that there are two steps involved in the evaluation of a partition function by this method:

(1) tabulate the contributions $\theta(t)$, beginning with the 'small' types t;

(2) for the given graph G, find $N(G, t)$ by counting subgraphs.

In practice, step (1) will be fairly simple, but step (2) is likely to be quite complicated, due especially to the difficulty of counting the disconnected types. It is not easy to evaluate N(G, t) when t is a type of graph with several components, each of which may be located anywhere on G.

In this chapter we shall show how to evaluate $\overline{Z}(\mathfrak{M}, G)$ for a more general class of models $\mathfrak{M}$, by counting only some of the connected types in G. This simplification of the counting problem leads to a corresponding loss - the modified contributions, replacing $\theta(t)$, are rather more esoteric. In order to set up the machinery, and give a precise definition of the word 'type', we require a little of the theory of separability in graphs.

A vertex v is said to be a **cut-vertex** of a graph G if, when v and its incident edges are removed from G, the number of components is thereby increased. A connected graph which has no cut-vertices is said to be **nonseparable**. A block of a graph is a maximal, nonseparable, edge-subgraph. These definitions are illustrated in Fig. 15. (In some cases it is convenient to regard isolated vertices as blocks, but this need not concern us unduly.)

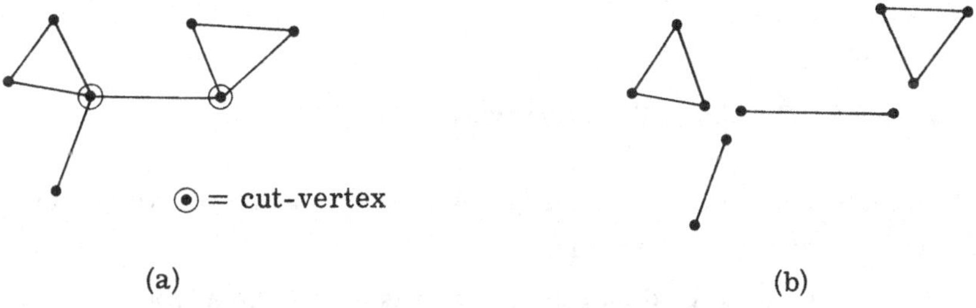

Fig. 15. A graph, its cut-vertices, and its blocks

We can now be more specific about the notion of 'type'. Two graphs G and G' are said to be **isomorphic** if there is a one-to-one correspondence $v \longleftrightarrow v'$ between their vertices, such that v and w are joined by an edge in G if and only if v' and w' are joined by an edge in G'. (See Fig. 16.) Two graphs are said to be **of the same type** if there is a one-to-one correspondence between their blocks, such that corresponding blocks are isomorphic. It must be emphasised that graphs of the same type are not necessarily isomorphic, although their blocks are isomorphic when taken in the correct order. The two diagrams in Fig. 17 represent graphs of the same type, but they are not isomorphic, since there are more vertices in the second graph than in the first. Of course,

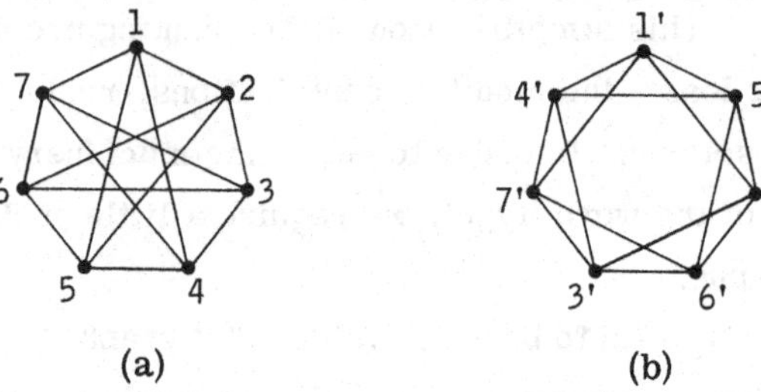

(a) (b)

Fig. 16. Isomorphic graphs

(a) (b)

Fig. 17. Graphs of the same type

if there is only one block - the nonseparable case - then 'isomorphic' and 'of the same type' are synonymous.

A function f defined on finite graphs is said to be a type-invariant if $f(G) = f(H)$ whenever G and H are graphs of the same type.

Theorem 7. Let $\mathfrak{M} = (A, i)$ be any interaction model. The function ξ defined by

$$\xi(G) = \overline{Z}(\mathfrak{M}, G)$$

is a type-invariant.

Proof. Suppose that the edge-set E of G is partitioned into two disjoint non-empty subsets E_1 and E_2, in such a way that the subgraphs $G_1 = \langle E_1 \rangle$ and $G_2 = \langle E_2 \rangle$ have either 0 or 1 common vertices. In other words, G_1 and G_2 are unions of blocks of G. We shall show that

$$\xi(G) = \xi(G_1)\xi(G_2). \tag{4.1.1}$$

Suppose first that G_1 and G_2 are disjoint. Then the set $\Omega = \Omega(G, A)$ of states ω on G is in one-to-one correspondence with the set $\Omega_1 \times \Omega_2$ of ordered pairs (ω_1, ω_2), where $\Omega_l = \Omega(G_l, A)$ for $l = 1, 2$. Now

$$Z(\mathfrak{M}, G_l) = \sum_{\omega_l} I_l(\omega_l),$$

where I_l is a product over E_l, and the sets E_1 and E_2 are disjoint. Hence

$$\begin{aligned} Z(\mathfrak{M}, G_1)Z(\mathfrak{M}, G_2) &= \Big(\sum_{\omega_1} I_1(\omega_1)\Big)\Big(\sum_{\omega_2} I_2(\omega_2)\Big) \\ &= \sum_{(\omega_1, \omega_2)} I_1(\omega_1)I_2(\omega_2) \\ &= \sum_{\omega} I(\omega) \\ &= Z(\mathfrak{M}, G). \end{aligned}$$

In this case $|VG| = |VG_1| + |VG_2|$, so that dividing through by $m^{|VG|}$ (where $m = |A|$ as usual) we obtain (4.1.1).

In the case when G_1 and G_2 have a common vertex v, Ω may be identified with the subset of $\Omega_1 \times \Omega_2$ consisting of those pairs (ω_1, ω_2) for which $\omega_1(v) = \omega_2(v)$. For any pair (θ_1, θ_2) we may define a state $(\theta_1, \overline{\theta}_2)$ in Ω, by setting

$$\overline{\theta}_2(u) = \theta_2(u) + \{\theta_1(v) - \theta_2(v)\} \qquad (u \in VG_2).$$

We notice that $\overline{\theta}_2(v) = \theta_1(v)$. Furthermore, there are m pairs (θ_1, θ_2) which give rise to a single $(\theta_1, \overline{\theta}_2)$, namely those for which the difference $\theta_2(u) - \overline{\theta}_2(u)$ is constant on all vertices u of G_2. For each such θ_2 we have $\delta\theta_2 = \delta\overline{\theta}_2$ and hence $I_2(\theta_2) = I_2(\overline{\theta}_2)$. Consequently, a calculation similar to that used in the first case shows that

$$Z(\mathfrak{M}, G_1)Z(\mathfrak{M}, G_2) = m\,Z(\mathfrak{M}, G),$$

and since we now have $|VG| = |VG_1| + |VG_2| - 1$, division by $m^{|VG|}$ again yields the result (4.1.1).

By repeated application of (4.1.1) we deduce that

$$\xi(G) = \prod_B \xi(B), \qquad (4.1.2)$$

where the product is over all blocks B of G. By its definition, ξ is an isomorphism invariant, and so in particular it is the same for two isomorphic blocks. The formula (4.1.2) thus implies that ξ is a type-invariant. //

4.2 The subgraph counting problem

Let St denote the set of isomorphism classes of nonseparable graphs (often called 'stars'). As we have already pointed out, in the nonseparable case an isomorphism class and a 'type' are the same thing. We shall use small Greek letters ρ, σ, τ, ... to denote these 'star types'; the star types with fewer than seven edges are depicted in Fig. 18.

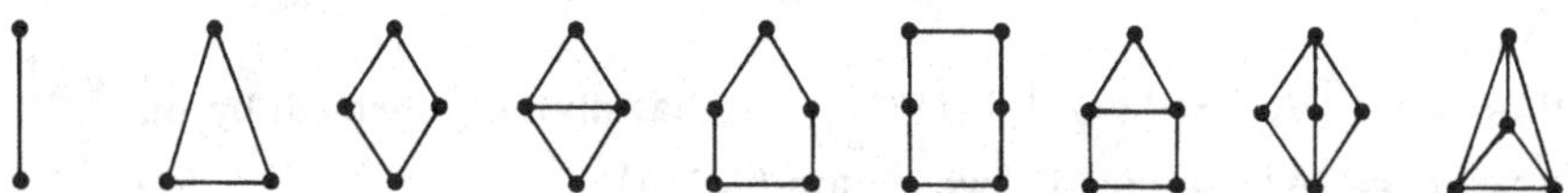

Fig. 18. Star types with < 7 edges

Any 'graph type' t may be regarded as a function defined on the set St and taking non-negative integer values; for each ρ in St the number $t(\rho)$ is the number of blocks of t with star type ρ. Accordingly, we may define the set Gr of **graph types** to be the set of functions

$$t : St \to \{0, 1, 2, 3, \ldots\},$$

with the property that $t(\rho)$ is nonzero for only a finite number of star types ρ. We often regard St itself as a subset of Gr, by identifying the star type σ with the graph type $\hat{\sigma}$ given by

$$\hat{\sigma}(\sigma) = 1; \quad \hat{\sigma}(\rho) = 0 \quad \text{for all } \rho \neq \sigma.$$

Note that the function o which always takes the value zero is a graph type but not a star type. Frequently, we shall use the symbols g, s, t, ... to denote graph types.

We denote by X and Y respectively the vector spaces of real-valued functions defined on St and Gr respectively. The vector space operations of addition and scalar multiplication are defined in the usual way:

$$(x_1 + x_2)(\rho) = x_1(\rho) + x_2(\rho); \quad (rx)(\rho) = r\,x(\rho),$$

for vectors x_1, x_2, x, in X and any real number r; and similarly in Y. Since St is to be regarded as a subset of Gr we have a linear mapping $J : Y \to X$ defined by

$$(Jy)(\rho) = y(\rho), \tag{4.2.1}$$

where ρ on the right-hand-side is really the graph type $\hat{\rho}$.

We proceed to express the subgraph counting problem, introduced in the previous section, in these terms. Let G be any graph, and let N(G, s) denote the number of subsets S of EG such that the edge-subgraph $\langle S \rangle$ is a graph of type s. It is a simple consequence of the definitions that, if G and H are graphs of the same type, then N(G, s) = N(H, s); in other words, N(, s) is a type-invariant for each fixed s. We introduce a set $\{c_g\}$ of vectors in Y, one for each graph type g, by setting

$$c_g(s) = N(G, s) \tag{4.2.2}$$

where G is any graph of type g. The vector c_g enumerates the set of all types of subgraph of a graph of type g.

Applying the mapping J, we see that Jc_g is a vector in **X**, and that it enumerates the set of all nonseparable subgraphs of a graph of type g. A fundamental result, originally due to Whitney, may be expressed in the following way: there is a nonlinear mapping $W : \mathbf{X} \to \mathbf{Y}$ such that

$$W(Jc_g) = c_g \quad \text{for all graph types } g.$$

Roughly speaking, the list c_g of all types of subgraph may be recovered if we know only the list Jc_g of nonseparable types. Furthermore, the procedure given by W is universal, in that it does not depend on g.

The definition of W is fairly straightforward. Let T be a given graph of type t, and let s be any graph type; we define B_{st} to be the number of ways of covering T with the blocks of a graph of type s. Precisely, suppose that the star types are ordered in some way: $\rho_1, \rho_2, \rho_3, \ldots$. Then B_{st} is the number of ordered families $\mathcal{S} = (S_1, S_2, \ldots, S_l)$, where each S_i is a non-empty subset of ET, and

(a) the union of the sets S_i is ET;

(b) each edge-subgraph $\langle S_i \rangle$ is nonseparable;

(c) the first $s(\rho_1)$ edge-subgraphs, $\langle S_1 \rangle, \ldots, \langle S_{s(\rho_1)} \rangle$, are of type ρ_1, the next $s(\rho_2)$ are of type ρ_2, and so on, so that l is the total number of blocks of s.

Suppose that the graph types are arranged in order, first by number of edges, and then, for types with the same number of edges, by number of blocks. If s comes before t in this ordering, then $B_{st} = 0$ since there are not enough edges (or blocks) for a covering of T by the blocks of s. Thus (B_{st}) is a lower triangular matrix, and, since each diagonal entry B_{ss} is nonzero, it is invertible. A small portion of this matrix is displayed on p. 67, with a (possibly) self-explanatory symbolism for graph types. Since we have agreed that there is a null graph type o, the matrix should really be augmented by another row and column, with $B_{oo} = 1$ and $B_{so} = B_{ot} = 0$ for all s and t except o. The matrix (B_{st}) induces a linear mapping $B : \mathbf{Y} \to \mathbf{Y}$, defined by

$$(By)(s) = \sum_t B_{st} y(t) . \qquad (4.2.3)$$

B:	I	II	△	III	□	I△	IIII	⊠	⬠	I□	II△	IIIII
I	1											
II	1	2										
△	0	0	1									
III	1	6	6	6								
□	0	0	0	0	1							
I△	0	0	3	0	0	1						
IIII	1	14	36	36	24	24	24					
⊠	0	0	0	0	0	0	0	1				
⬠	0	0	0	0	0	0	0	0	1			
I□	0	0	0	0	4	0	0	1	0	1		
II△	0	0	9	0	0	7	0	4	0	0	2	
IIIII	1	30	150	240	240	240	240	120	120	120	120	120

The sum is over all graph types t, but it is essentially finite, since the matrix (B_{st}) is lower triangular.

We shall also require the nonlinear mapping $U : X \to Y$ defined by

$$(Ux)(t) = \prod_{\rho} \{x(\rho)\}^{t(\rho)} . \qquad (4.2.4)$$

The product is over all star types ρ, but once again it is essentially finite: this is because $t(\rho) \neq 0$ only for finitely many ρ, and so all but a finite number of factors are equal to 1.

Theorem 8. Let the vectors c_g and the mappings

$J : Y \to X$; $B : Y \to Y$; $U : X \to Y$;

be as defined above. Then writing $W = B^{-1}U$ we have

$$W(Jc_g) = c_g \text{ for all } g \in Gr. \qquad (4.2.5)$$

Proof. We have already remarked that B is invertible; thus it is sufficient to show that $UJc_g = Bc_g$. Let G be any graph of type g, and consider

$$(Bc_g)(s) = \sum_t B_{st} c_g(t).$$

Given an edge-subgraph of G, of type t, there are B_{st} ways of covering it with the blocks of a graph of type s. Hence $B_{st}c_g(t)$ is just the number of ways of imbedding the blocks of s into G, so that the union is a graph of type t. The sum over all t is just the total number of ways of imbedding the blocks of s into G. To be precise, it is the number of ordered families $\mathcal{S}$ of non-empty subsets of EG, satisfying conditions (b) and (c) on page 66.

Now, we may compute this number in a different way. For each star type ρ, there are $s(\rho)$ blocks of type ρ to be imbedded, and G has $Jc_g(\rho)$ edge-subgraphs of this type. So these blocks can be imbedded in $\{Jc_g(\rho)\}^{s(\rho)}$ ways, and the whole set of blocks in

$$\prod_\rho \{Jc_g(\rho)\}^{s(\rho)} = (UJc_g)(s)$$

ways. Hence $Bc_g = UJc_g$ as required. //

A few explicit calculations will help to clarify the meaning of Theorem 8. We shall obviously need the inverse of B, which may be computed recursively (see page 69). An element x of X may be written as a column vector with components $(x_{|}, x_{\triangle}, x_{\square}, \ldots)$ where, for example, $x_{\triangle}$ denotes the value of x on the triangle star type $\triangle$. For any graph type t, (Ux)(t) is a monomial expression in $x_{|}, x_{\triangle}, x_{\square}, \ldots$; for example

$$(Ux)(\,||\triangle\,) = x_{|}^2\, x_{\triangle}\,.$$

We see that

$$(Wx)(s) = (B^{-1}Ux)(s) = \sum_t B^{-1}_{st}(Ux)(t)$$

is a polynomial expression $\phi_s(x_{|}, x_{\triangle}, x_{\square}, \ldots)$, with coefficients given by the relevant row of B^{-1}. For instance

B^{-1}:	I	II	△	III	□	I△	IIII	⊠	⬠	I□	II△	IIIII
I	1											
II	$-\frac{1}{2}$	$\frac{1}{2}$										
△	0	0	1									
III	$\frac{1}{3}$	$-\frac{1}{2}$	-1	$\frac{1}{6}$								
□	0	0	0	0	1							
I△	0	0	-3	0	0	1						
IIII	$-\frac{1}{4}$	$\frac{11}{24}$	3	$-\frac{1}{4}$	-1	-1	$\frac{1}{24}$					
⊠	0	0	0	0	0	0	0	1				
⬠	0	0	0	0	0	0	0	0	1			
I□	0	0	0	0	-4	0	0	-1	0	1		
II△	0	0	6	0	0	$-\frac{7}{2}$	0	-2	0	0	$\frac{1}{2}$	
IIIII	$\frac{1}{5}$	$-\frac{5}{12}$	-6	$\frac{7}{24}$	4	$\frac{7}{2}$	$-\frac{1}{12}$	2	-1	-1	$-\frac{1}{2}$	$\frac{1}{120}$

$$\phi_{|\triangle}(x_{|}, x_{\triangle}, \ldots) = -3x_{\triangle} + x_{|}\, x_{\triangle}\,;$$

$$\phi_{||\triangle}(x_{|}, x_{\triangle}, \ldots) = 6x_{\triangle} - \tfrac{7}{2}x_{|}\, x_{\triangle} - 2x_{\boxtimes} + \tfrac{1}{2}x_{|}^{2}\, x_{\triangle}\,.$$

If any graph G is given, and the variables x_ρ are set equal to the numbers of edge-subgraphs of G having star type ρ, then the value of $\phi_s(x_{|}, x_{\triangle}, x_{\square}, \ldots)$ is the number of edge-subgraphs of G which have graph type s. This is the essential content of Whitney's theorem. The formulae for ϕ_s may be checked directly in simple cases, such as those given above. For example, the number of subgraphs of type I△ is equal to the number $(x_{\triangle})$ of ways of picking a triangle, multiplied by the number $(x_{|} - 3)$ of ways of picking one of the remaining edges.

4.3 The cluster expansion

We now investigate methods of evaluating $\overline{Z}(\mathfrak{M}, G)$ which require merely the counting of the nonseparable subgraphs of G, rather than all subgraphs. The underlying idea is to regard graphs as vectors in the space **X**, each graph G of type g being represented by the vector Jc_g which enumerates its nonseparable subgraphs. The logarithm of the reduced partition function then becomes a linear functional on **X**.

It is convenient to work in a finite-dimensional subspace of **X**. Let St_N denote the set of star types with not more than N edges, and let $\mathbf{X}_N$ denote the subspace of **X** consisting of those functions on St which vanish identically outside St_N. We say that a vector $\mathbf{x}_0$ has **finite support** if it belongs to some $\mathbf{X}_N$. The set of vectors

$$Jc_\gamma \quad (\gamma \in St_N)$$

is a basis for $\mathbf{X}_N$. To prove this, we need only show that the set is linearly independent, since it has the right cardinality for a basis. Now $(Jc_\sigma)(\sigma) = 1$, while $(Jc_\tau)(\sigma) = 0$ for any star type τ which has not more edges than σ has; hence Jc_σ is not linearly dependent on the set of such vectors Jc_τ. It follows that the whole set is linearly independent.

Suppose that $\mathcal{P}$ is a positive interaction model. Then $\overline{Z}(\mathcal{P}, G)$ is positive, and its logarithm $\ln \overline{Z}(\mathcal{P}, G)$ is properly defined. We introduce a linear functional $\mathcal{L}$ on the space $\mathbf{X}_N$ by assigning its values on the basis $\{Jc_\gamma\}$ as follows:

$$\mathcal{L}(Jc_\gamma) = \ln \overline{Z}(\mathcal{P}, \Gamma), \tag{4.3.1}$$

where Γ is any nonseparable graph of star type γ. Any vector $\mathbf{x}$ in $\mathbf{X}_N$ may be expressed uniquely as a linear combination of basis vectors; if $\mathbf{x} = \sum_\gamma \xi_\gamma (Jc_\gamma)$, then we set

$$\mathcal{L}(\mathbf{x}) = \sum_\gamma \xi_\gamma \mathcal{L}(Jc_\gamma). \tag{4.3.2}$$

We must check that, with this definition of $\mathcal{L}(\mathbf{x})$, the vectors Jc_g correctly represent graphs G, where g is any graph type. In other words, we must show that

$$\mathcal{L}(Jc_g) = \ln \overline{Z}(\mathcal{P}, G) \tag{4.3.3}$$

for any graph G of type g. Now any subgraph of star type σ must be contained in a single block of G, and so we have the identity

$$(Jc_g)(\sigma) = \sum_\beta g(\beta)(Jc_\beta)(\sigma) ,$$

or

$$Jc_g = \sum_\beta g(\beta)Jc_\beta .$$

Substituting Jc_g for x in (4. 3. 2) and using (4. 3. 1) we obtain

$$\begin{aligned}\mathcal{L}(Jc_g) &= \sum g(\beta)\mathcal{L}(Jc_\beta)\\ &= \sum g(\beta) \ln \overline{Z}(\mathcal{P}, B),\end{aligned}$$

where B is a block of type β. The right-hand-side is just $\ln \overline{Z}(\mathcal{P}, G)$, by the formula (4. 1. 2) derived in the proof of the fact that $\overline{Z}$ is a type-invariant. Hence we have the result.

Theorem 9. Let $\mathcal{P}$ be any positive interaction model. There is a vector π in X such that for any graph G we have

$$\ln \overline{Z}(\mathcal{P}, G) = \sum_\rho N(G, \rho)\pi(\rho) , \tag{4. 3. 4}$$

where $N(G, \rho)$ is the number of subgraphs of G which have star type ρ.

Proof. For each star type ρ, define an element e_ρ of X as follows:

$$e_\rho(\sigma) = \begin{cases} 1 & \text{if } \sigma = \rho; \\ 0 & \text{otherwise.}\end{cases}$$

Clearly, the set $\{e_\rho\}$ (where ρ runs through the set St_N) is a basis for X_N, and for any x in X_N we have

$$x = \sum_\rho x(\rho)e_\rho .$$

Since $\mathcal{L}$ is a linear functional on X_N,

$$\mathcal{L}(x) = \sum_\rho x(\rho)\mathcal{L}(e_\rho) .$$

Suppose that the graph G has type g. Substituting Jc_g for x in the preceding formula we obtain

$$\mathcal{L}(Jc_g) = \sum_\rho (Jc_g)(\rho)\mathcal{L}(e_\rho) .$$

The left-hand-side has been shown (4. 3. 3) to be equal to $\ln \overline{Z}(\mathcal{P}, G)$, and $(Jc_g)(\rho) = N(G, \rho)$; so defining $\pi(\rho)$ to be $\mathcal{L}(e_\rho)$ we have the required formula:

$$\ln \overline{Z}(\mathcal{P}, G) = \sum_\rho N(G, \rho)\pi(\rho) . \quad //$$

The formula contained in Theorem 9 is of the kind known to theoretical physicists as a 'cluster expansion'. It involves only the nonseparable subgraphs of G, and in that respect it is a considerable improvement on the expansions of resonant models obtained in Section 2. 1. However, there remains the problem of determining the mysterious π vector.

In one sense, the problem is trivial, since π is given by the values of the linear functional $\mathcal{L}$ on the basis $\{e_\rho\}$. Now we may express each e_ρ in terms of the original basis $\{Jc_\gamma\}$ - this amounts to inverting the matrix whose columns are the vectors $\{Jc_\gamma\}$. The obvious triangular property of the conversion matrix means that we can build up a table of values of $\pi(\rho) = \mathcal{L}(e_\rho)$ by evaluating $\mathcal{L}(Jc_\gamma) = \ln \overline{Z}(\mathcal{P}, \Gamma)$ for star types γ which are not 'larger' than ρ. But since our main problem is the evaluation of the partition function, there is an element of circularity involved in this procedure.

However, the theory can be carried further so that it links up with the simple expansion techniques of Section 2. 1. This development leads to an alternative method of finding π.

4. 4 Subgraph expansions revisited

We introduce inner products $\langle\!\langle , \rangle\!\rangle$ and $\langle , \rangle$ on the spaces X and Y respectively, defined by the formulae

$$\langle\!\langle x_1, x_2 \rangle\!\rangle = \sum_\rho x_1(\rho)x_2(\rho)$$

$$\langle y_1, y_2 \rangle = \sum_t y_1(t)y_2(t) .$$

In both cases the sum may be infinite, and so some restrictions on the vectors are necessary, but we shall deal with this difficulty as it arises. With this notation, the result of Theorem 9 is that

$$\overline{Z}(\mathcal{P}, G) = \exp\langle\!\langle \pi, Jc_g \rangle\!\rangle . \tag{4.4.1}$$

Suppose that any vectors x, x_o are given, where x is in X and x_o is in some subspace X_N. Since x_o has finite support, $\langle\!\langle x, x_o \rangle\!\rangle$ is a finite sum, and we have

$$\begin{aligned}\exp\langle\!\langle x, x_o \rangle\!\rangle &= \exp \sum_\rho x(\rho)x_o(\rho) \\ &= \prod_\rho \exp\{x(\rho)x_o(\rho)\} \\ &= \prod_\rho \sum_{n\geq 0} \frac{\{x(\rho)x_o(\rho)\}^n}{n!} .\end{aligned}$$

Here the product is over the finite set St_N and so we may apply the infinite form of the distributive law (Lemma II of Appendix A). We obtain

$$\exp\langle\!\langle x, x_o \rangle\!\rangle = \sum_t \prod_\rho \frac{\{x(\rho)x_o(\rho)\}^{t(\rho)}}{t(\rho)!} ,$$

where the sum is over all functions $t : St_N \to \{0, 1, 2, \dots\}$. But the sum may be equally taken over all graph types, since if $t(\sigma) \neq 0$ for some $\sigma \notin St_N$, then $x_o(\sigma) = 0$ and the summand for t is zero. The right-hand-side is an inner product (which we have proved to exist) in the space Y. If we introduce the mapping $V : X \to Y$ defined by

$$(Vx)(t) = (1/t!)\,(Ux)(t),$$

where $t!$ stands for the product of the factorials $t(\rho)!$ (which is the diagonal entry B_{tt} of the matrix B), then we have shown that $\langle Vx, Ux_o \rangle$ exists when x_o has finite support, and

$$\exp\langle\!\langle x, x_o \rangle\!\rangle = \langle Vx, Ux_o \rangle . \tag{4.4.2}$$

This equation is the vital link between cluster expansions and the ordinary expansions of the kind introduced in Chapter 2.

Theorem 10. For any interaction model $\mathfrak{M}$ there is a unique vector m in Y such that

$$\overline{Z}(\mathfrak{M}, G) = \sum_t N(G, t)m(t) . \tag{4.4.3}$$

If the model is positive, then the vector π whose existence is asserted in Theorem 9 is given by

$$\pi = J(B^{-1})'m . \tag{4.4.4}$$

Proof. The proof of the first part is very similar to the proof of Theorem 9. Let Gr_N denote the set of graph types with at most N edges, and Y_N the subspace of Y consisting of those vectors which vanish outside Gr_N. Then the set of vectors c_g $(g \in Gr_N)$ is a basis for Y_N. Accordingly, we may define a linear functional $\mathcal{K}$ on Y_N by assigning its values on the basis, as follows:

$$\mathcal{K}(c_g) = \overline{Z}(\mathfrak{M}, G).$$

$\mathcal{K}$ is extended, by linearity, to the whole of Y_N. An alternative basis for Y_N is the set of vectors e_t $(t \in Gr_N)$ defined by $e_t(s) = 1$ if $s = t$, $e_t(s) = 0$ otherwise. Since $c_g = \Sigma c_g(t)e_t$ we have

$$\mathcal{K}(c_g) = \sum c_g(t)\mathcal{K}(e_t);$$

that is, writing $m(t)$ for $\mathcal{K}(e_t)$,

$$\overline{Z}(\mathfrak{M}, G) = \sum_t N(G, t)m(t) .$$

Furthermore, the change of basis from $\{c_g\}$ to $\{e_t\}$ is an invertible mapping, and so $m(t) = \mathcal{K}(e_t)$ is uniquely determined by the values $\mathcal{K}(c_g) = \overline{Z}(\mathfrak{M}, G)$.

For the second part, suppose that π is as given in Theorem 9, so that from (4.4.1) and (4.4.2) we deduce

$$\overline{Z}(\mathfrak{M}, G) = \exp\langle\!\langle \pi, Jc_g \rangle\!\rangle = \langle V\pi, UJc_g \rangle . \tag{4.4.5}$$

We shall need to know that $B'V\pi$ exists, where B' is the transpose of B. This is not obvious, since $(B'V\pi)(t)$ is defined formally by the infinite series

$$\sum_s B_{st}(V\pi)(s) .$$

It is clearly sufficient to prove the convergence of this series when π is positive. Now

$$B_{st} \leq \sum_r B_{sr} c_t(r) = (Bc_t)(s) ,$$

since B_{st} is equal to one summand (the one with $r = t$) on the right-hand-side. But, from the proof of Theorem 8, $Bc_t = UJc_t$, and so

$$B_{st} \leq (UJc_t)(s) .$$

Hence the series for $(B'V\pi)(t)$ is majorised by

$$\sum_s (UJc_t)(s)(V\pi)(s) = \langle V\pi, UJc_t \rangle ,$$

which exists by the argument leading to (4.4.2) (since Jc_t has finite support). Hence $(B'V\pi)(t)$ exists for all graph types t.

Since $(Wx)(t) = (B^{-1}Ux)(t)$ also exists for all t, we may proceed as follows:

$$\begin{aligned}
\langle V\pi, Ux_o \rangle &= \sum_t V\pi(t)Ux_o(t) \\
&= \sum_{s,t} V\pi(s)Ux_o(t)(BB^{-1})_{st} \\
&= \sum_{s,t} \sum_r B_{sr} V\pi(s)(B^{-1})_{rt} Ux_o(t) \\
&= \sum_r (B'V\pi)(r)(B^{-1}Ux_o)(r) \\
&= \langle B'V\pi, B^{-1}Ux_o \rangle \\
&= \langle B'V\pi, Wx_o \rangle .
\end{aligned}$$

Hence, putting $x_o = Jc_g$, we have

$$\begin{aligned}
\overline{Z}(\mathfrak{M}, G) &= \langle V\pi, UJc_g \rangle \quad \text{from (4.4.5)} \\
&= \langle B'V\pi, WJc_g \rangle \\
&= \langle B'V\pi, c_g \rangle \quad \text{from Theorem 8.}
\end{aligned}$$

But the first part of this theorem asserts that there is a unique m such that

$$\overline{Z}(\mathfrak{M}, G) = \sum_t N(G, t)m(t) = \langle m, c_g \rangle .$$

Consequently, we must have $m = B'V\pi$ and $J(B^{-1})'m = \pi$, as required. //

The first part of Theorem 10 is a generalization of the results obtained in Section 2.1. For example, if we consider a positive resonant model $\mathcal{R}$, normalized so that $i_0 > 0$ and $i_1 = 1$, then equation (2.1.9) becomes

$$\overline{Z}(\mathcal{R}, G) = \sum_{S \subseteq EG} (i_0 - 1)^{|S|}(1/m)^{r\langle S \rangle} .$$

Putting $x = i_0 - 1$, and collecting terms according to the type of $\langle S \rangle$, we obtain

$$\overline{Z}(\mathcal{R}, G) = \sum_t N(G, t)x^{e(t)}(1/m)^{r(t)} ,$$

where $e(t)$ and $r(t)$ denote the number of edges and rank of t. Thus the unique vector m associated with this model is given by the simple formula

$$m(t) = x^{e(t)}(1/m)^{r(t)} .$$

Furthermore, the π vector occurring in the cluster expansion of $\mathcal{R}$ is determined by the relation $\pi = J(B^{-1})'m$.

The general form of this relation is, explicitly,

$$\pi(\rho) = \sum_t (B^{-1})_{t\rho} m(t) .$$

In other words, $\pi(\rho)$ is a sum of terms $m(t)$, with coefficients taken from column ρ of B^{-1}. The first non-zero entry in this column is $(B^{-1})_{\rho\rho} = 1$, and so

$$\pi(\rho) = m(\rho) + \sum{}^*(B^{-1})_{t\rho} m(t) ,$$

where Σ^* denotes a sum over types t which come after ρ in the ordering. Roughly speaking, $\pi(\rho)$ is of the same order of magnitude as $m(\rho)$.

This useful fact underlies the 'Theorem of Vanishing Coefficients', which can be used to justify a recursive method for evaluating rank polynomials.

4.5 Vertex-transitive graphs

An automorphism of a graph G is a one-to-one correspondence (permutation) $\alpha : VG \to VG$, with the property that two vertices v and w are joined by an edge of G if and only if $\alpha(v)$ and $\alpha(w)$ are also joined by an edge. For example, the permutation $\alpha = (01234)(56789)$ is an automorphism of the graph depicted in Fig. 19. The automorphisms

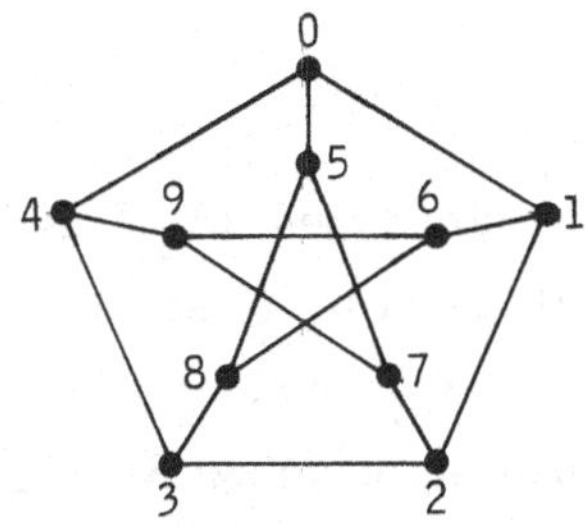

Fig. 19. A vertex-transitive graph

of a graph may be combined by the usual method of composing permutations, and the set of all automorphisms of G thereby becomes a group. The graph G is said to be **vertex-transitive** if, for any two vertices x and y there is some automorphism of G which takes x to y. The graph in Fig. 19 has this property, as may be seen by combining suitable powers of the automorphism α with the automorphism $\beta = (05)(18)(23)(47)$. Clearly, in a vertex-transitive graph all vertices are alike with respect to graph properties - this fact alone is sufficient for our present purposes.

The theory of cluster expansions takes an especially simple form when applied to vertex-transitive graphs. We begin by translating the formula (4.4.1),

$$\overline{Z}(\mathcal{P}, G) = \exp\langle\!\langle \pi, Jc_g \rangle\!\rangle,$$

into the language of 'vertex-subgraphs'. Each subset V' of VG defines a **vertex-subgraph** $\langle V' \rangle$: the vertex-set of $\langle V' \rangle$ is just V', and its edge-set consists of those edges of G which have both ends in V'. For any two star types σ and τ, define

$A_{\sigma\tau}$ = the number of spanning edge-subgraphs of a graph of type τ which have star type σ.

The adjective 'spanning' indicates that the relevant edge-subgraphs must contain all the vertices of τ; in other words, they are obtained by deleting some edges, but no vertices, from τ. Let $A : \mathbf{X} \to \mathbf{X}$ be the linear mapping induced by the matrix $(A_{\sigma\tau})$, that is,

$$A\mathbf{x}(\rho) = \sum_{\sigma} A_{\rho\sigma}\mathbf{x}(\sigma) .$$

With a suitable ordering of the star types $(A_{\sigma\tau})$ is lower triangular, and its diagonal entries are all equal to 1, so A is invertible. Furthermore, each row and column of $(A_{\sigma\tau})$ contains only finitely many non-zero entries, so that the transpose mapping A' is properly defined, as is $(A')^{-1}$. We shall show that

$$(A')^{-1}(Jc_g) = g ,$$

where $g(\sigma)$ denotes the number of vertex-subgraphs of G which have star type σ.

To prove this, we remark that each edge-subgraph of G, of star type τ, is a spanning edge-subgraph of just one non-separable vertex-subgraph of G. Hence we have the equation

$$Jc_g(\tau) = \sum_{\sigma} A_{\sigma\tau} g(\sigma) ,$$

or

$$Jc_g = A'g .$$

Inverting A' gives the result.

Let $p = A\pi$; then a simple calculation shows that

$$\langle\!\langle \pi, Jc_g \rangle\!\rangle = \langle\!\langle A\pi, (A')^{-1} Jc_g \rangle\!\rangle = \langle\!\langle p, g \rangle\!\rangle ,$$

so that

$$\overline{Z}(\mathcal{P}, G) = \exp\langle p, g\rangle . \qquad (4.5.1)$$

Formula (4. 5. 1) is analogous to (4. 4. 1), but it involves vertex-subgraphs rather than edge-subgraphs. Incidentally, we note that g has finite support, since Jc_g does.

When G is vertex-transitive, there is a simplified form of (4. 5. 1) which has some importance. Let z_0 be any given vertex of G, and define a vector g_0 in $\mathbf{X}$ by

$$g_0(\sigma) = \text{number of vertex-subgraphs of } G \text{ which have type } \sigma \text{ and contain } z_0.$$

Since G is vertex-transitive, g_0 is the same, whatever vertex of G we choose for z_0. Now there are two ways of counting the pairs (z, Z) such that $z \in Z \subseteq VG$ and $\langle Z\rangle$ has type σ. Counting first the vertices z in Z, we obtain the total $|Z|\, g(\sigma)$, while counting first the subsets Z containing a given z, we obtain $g_0(\sigma)|VG|$. Hence, if $v(\sigma) = |Z|$,

$$v(\sigma)g(\sigma) = |VG|g_0(\sigma). \qquad (4.5.2)$$

Putting $p_0(\sigma) = p(\sigma)/v(\sigma)$, and substituting in (4. 5. 1), we obtain

$$\overline{Z}(\mathcal{P}, G) = \exp\langle p_0, |VG|g_0\rangle ,$$

or

$$\overline{Z}(\mathcal{P}, G)^{1/|VG|} = \exp\langle p_0, g_0\rangle . \qquad (4.5.3)$$

The practical significance of (4. 5. 3) is that the subgraphs to be counted have finally been 'pinned down'. In order to evaluate g_0 we need only count star subgraphs containing a specified vertex z_0. If we wish to check that we have located all such subgraphs, with at most n vertices, then we can confine our search to a neighbourhood of z_0, with diameter n.

The formula (4. 5. 3) also has a theoretical significance. If we have an infinite vertex-transitive graph G^∞, then it has a g_0 vector defined precisely as for a finite graph, since each $g_0(\sigma)$ will be finite, provided only that G^∞ has finite valency. The sole difference from the finite case is that g_0 no longer has finite support; nevertheless, we might hope to define the reduced partition function for an infinite vertex-transitive graph by the formula

$$\overline{Z}_\infty(\mathcal{P}, G^\infty) = \exp\langle\langle p_0, g_0\rangle\rangle. \qquad (4.5.4)$$

The relevant question is: for a given model $\mathcal{P}$, with associated vector p_0, what constraints on x are sufficient for the existence of the inner product $\langle\langle p_0, x\rangle\rangle$?

NOTES AND REFERENCES FOR CHAPTER 4

Whitney's original treatment [3] of the subgraph counting problem was rather obscure, although it is possible to discern the basic features of the present method. One complication seems to stem from the fact that arguments about the matrix B are expressed in terms of B^{-1}. The same paper also contains the essence of the method of cluster expansions. However, the development of these expansions as such is due to various workers in theoretical physics: see the references in Domb [1]. Tutte [2] independently derived the results for the general rank polynomial, which corresponds in our terminology to the case of a resonant model. He also gave a proof of the 'Theorem of Vanishing Coefficients'. An explanation of the importance of this theorem in the calculation of chromatic polynomials may be found in Chapter 12 of the author's book (reference [1] of Chapter 1).

1. Domb, C. Graph theory and embeddings. Phase Transitions and Critical Phenomena (C. Domb and M. S. Green, eds.), Volume 3, pp. 1-96.
2. Tutte, W. T. On dichromatic polynomials. J. Combinatorial Theory, 2 (1967), 301-20.
3. Whitney, H. The colouring of graphs. Ann. Math. 3 (1932), 688-718.

5 · Prospects

5.1 Symmetry and dimensionality

This final chapter should be the culmination of the earlier ones, but it is not. The present state of our knowledge is most unsatisfactory, and it is not possible to present neat and final conclusions. Accordingly, the contents of this chapter are speculative: we shall cover a few topics which may have some part to play in the theory, when it becomes clear what that theory should be.

The first topic is concerned with the ideas of symmetry and dimensionality. Let us begin by recalling the notion of a graph scheme $G = (F, J)$, as discussed in Chapter 2. Associated with such a scheme we have a sequence $\{G_n\}$ of finite graphs, where G_n consists of n copies of the fundamental graph F, linked cyclically by edges as prescribed by the set J. Specifically, for each $l = 1, 2, \ldots, n$, and each vertex v and edge e of F, there is one vertex v_l and one edge e_l of G_n; in addition there are edges j_l corresponding to each j in J. The 'limit' of the sequence $\{G_n\}$ may be thought of as an infinite frieze, with copies of F linked in the same way. In fact, we may define the infinite graph $L = \text{Lim } G$ as follows.

Let I denote the set of integers and for each i in I let

$$V^{(i)} = \{v_i | v \in VF\}, \quad E^{(i)} = \{e_i | e \in EF\} \cup \{j_i | j \in J\}.$$

Then the vertex-set of L is the union of all the $V^{(i)}$, and its edge-set is the union of all $E^{(i)}$; the vertices v_i are incident with edges e_i as in F, and there are edges j_i joining v_i to w_{i+1} for each $j = (u, w)$. In a sense, L is the simplest kind of infinite graph, since it is basically one-dimensional. Our aim here is to provide (in this context) a formal definition of the dimensionality of an infinite graph, and to compare the properties of interaction models in the one-dimensional and two-dimensional cases.

The notion of symmetry is crucial. In Chapter 4 we mentioned that the set of all automorphisms of a given graph G is a group. Here we shall be concerned with the existence of certain subgroups of this group, which reflect certain symmetry properties of the graph. If a set Γ of automorphisms of G is a group (a subgroup of the full automorphism group), then we say that Γ **acts on** G.

We shall be concerned mainly with the action of cyclic groups. A group is said to be a **finite cyclic group of order** n if it contains an element ξ such that every element of the group is a power of ξ, and $\xi^n = \varepsilon$, the identity element. We shall use the symbol Z_n to denote any group of this kind:

$$Z_n = \{\varepsilon, \xi, \xi^2, \ldots, \xi^{n-1}\},$$

and we say that Z_n is **generated by** ξ. Referring to Fig. 19 (p. 77), we see that the permutation $\alpha = (01234)(56789)$ generates a cyclic group Z_5 which acts on the graph depicted in that figure. Slightly less obvious is the fact that the group $Z_3 = \{\varepsilon, \theta, \theta^2\}$ acts on the same graph, where $\theta = (0)(145)(298)(376)$.

A group is said to be an **infinite cyclic group** if it contains an element ξ, together with all its powers and their inverses, and no other elements. We write

$$Z = \{\ldots, \xi^{-2}, \xi^{-1}, \varepsilon, \xi, \xi^2, \ldots\}.$$

For example, a group of this kind acts on the infinite linear chain, where ξ is an automorphism which takes each vertex to its right-hand neighbour.

Our only other group-theoretic requirement is the notion of the direct product of cyclic groups. A group is said to be a **direct product** $Z_r \times Z_s$ if it contains two elements ξ and η such that $\xi^r = \varepsilon$, $\eta^s = \varepsilon$, and $\xi\eta = \eta\xi$, and it is the smallest such group. The elements of a group $Z_r \times Z_s$ are the rs terms

$$\xi^i \eta^j \qquad (0 \leq i \leq r-1,\ 0 \leq j \leq s-1).$$

The **direct product** $Z \times Z$ is defined similarly: it contains all the terms $\xi^i \eta^j$, where i and j are any integers, and $\xi\eta = \eta\xi$. This may be extended in the obvious way to the case where there are more than two groups, and we shall write Z^d for the direct product of d infinite cyclic groups.

In the case of a graph scheme G, the group Z_n acts on the finite graph G_n, the generator ξ being defined by $\xi(v_l) = v_{l+1}$. The group Z acts on the infinite graph Lim G in the same way. In order to extend this to higher dimensions we must examine the nature of the actions rather more closely. First, we notice that the equation $\gamma(v) = v$ has a solution only when γ is the identity; in other words, the action has no **fixed vertex.** Secondly, the fundamental graph F, whose repetitions cover the vertex-set of the whole graph, is finite: this is expressed by saying that the group action has a finite number of 'orbits'. The **orbit** $O(v)$ of a vertex v in the action of a group Γ on a graph G is the set of images of the vertex v:

$$O(v) = \{w \in VG \mid w = \gamma(v) \text{ for some } \gamma \text{ in } \Gamma\}.$$

It is easy to see that two orbits $O(x)$ and $O(y)$ are either disjoint (when $x \notin O(y)$) or identical (when $x \in O(y)$). Hence the orbits partition VG. In the case of a cyclic group acting on the graphs arising from a scheme, there is one orbit for each vertex of the fundamental graph F.

We are now able to give the promised definition of the dimensionality of an infinite graph. (Since the term 'dimension' has been used elsewhere - in several different ways - we must employ the rather clumsy word 'dimensionality'.) Let K be a connected infinite graph, in which each vertex has finite valency. We say that K is **d-dimensional** if d is the smallest integer for which a group Z^d acts on K in such a way that

(i) there are no fixed vertices, and

(ii) the number of orbits is finite.

Roughly speaking, a d-dimensional graph K may be thought of as the infinite repetition, in d independent directions, of a finite graph, with certain links between the copies. We remark that the full automorphism group of K may be strictly larger than Z^d - for example, it may contain 'rotations' and 'reflections'.

We began by proposing to generalize the one-dimensional nature of the limit of a graph scheme. So our first task now is to formulate the exact relationship between a one-dimensional graph, as just defined, and graph schemes.

Suppose that the group Z, with generator ξ, acts on K in such a way that conditions (i) and (ii) are satisfied. Since there are a finite number of orbits, we may choose a finite set S of vertices, containing one vertex from each orbit. We claim that each vertex v of K has a unique representation in the form $v = \xi^i(s)$ for some integer i and some s in S. To prove this, suppose that

$$v = \xi^i(s) = \xi^j(s') .$$

Then $s' = \xi^{i-j}(s)$ and so s and s' are in the same orbit; since we have chosen just one vertex from each orbit, $s = s'$. Hence $s = \xi^{i-j}(s)$ and since there are no fixed vertices, we must have $i = j$.

We shall use the notation S^l for the set of vertices $\xi^l(s)$, where s belongs to S. Since K is connected and each vertex has finite valency, there is a largest positive integer b for which some vertex in S is joined to a vertex in S^b. Let

$$X = S \cup S^1 \cup S^2 \ \ldots \cup S^{b-1}, \quad \kappa = \xi^b;$$

in particular, if $b = 1$ then $X = S$ and $\kappa = \xi$. The definition of b ensures that the edges of K are of just two kinds: those that have both ends in the same iterate $\kappa^l(X)$, and those that have one end in $\kappa^l(X)$ and one end in $\kappa^{l+1}(X)$. Consequently, we may define a graph scheme $\mathbf{K} = (F, J)$ by setting $F = \langle X \rangle$ (the vertex-subgraph of K induced by X) and

$$J = \{(x, y) \in X \times X \,|\, x \text{ is joined to } \kappa^b(y) \text{ in } K\}.$$

Furthermore, the given graph K is the limit, Lim **K**, of this scheme.

We have proved that, given any one-dimensional graph K, there is a graph scheme **K** such that K = Lim **K**. The finite graphs K_n

associated with this graph scheme may be regarded as an approximating sequence for K. It is worth noting that the canonical action of **Z** (generated by κ) on Lim K, is not necessarily the same as the given action of **Z** (generated by ξ) on K. In fact, the first group **Z** is a subgroup of the second one. Of course, any infinite cyclic group possesses subgroups with the same structure, generated by powers of the given generator. The fact that **Z** (or $\mathbf{Z}^d$) has infinitely many subgroups of the same kind is the basis of the method of 'renormalization'.

Let us turn to the two-dimensional case, which is typical of all cases with $d \geq 2$.

Suppose that $\mathbf{Z}^2$ acts on K in such a way that conditions (i) and (ii) are satisfied. By following the method used in the one-dimensional case, with due attention to detail, we may construct a sequence $\{K_{(r,s)}\}$ of finite graphs which serves as an approximating sequence for K. More precisely, we set up a sequence of graph schemes

$$K(r) = (F(r),\ J(r))$$

in which $\mathbf{Z}_r$ acts on F(r) in a manner derived from the action of the first factor of $\mathbf{Z} \times \mathbf{Z}$ on K. The sth cyclic graph associated with the scheme K(r) is acted on by $\mathbf{Z}_s$, corresponding to the action of the second factor of $\mathbf{Z} \times \mathbf{Z}$ on K. We let this be our graph $K_{(r,s)}$ - that is,

$$K_{(r,s)} = (K(r))_s .$$

There is an action of $\mathbf{Z}_r \times \mathbf{Z}_s$ on $K_{(r,s)}$ compatible with an action of $\mathbf{Z} \times \mathbf{Z}$ on K, although (as in the one-dimensional case) the acting group may be a subgroup of the given $\mathbf{Z} \times \mathbf{Z}$.

This construction is illustrated for the simplest case, the plane square lattice graph ⊞ , in Fig. 20. In the associated schemes ⊞(r), the graph F(r) is simply a circuit graph C_r and the joins J(r) link corresponding vertices. It will be apparent that the graph $⊞_{(r,s)}$ is a 'toroidal' square lattice. The general two-dimensional case is similar, but each vertex of ⊞ is replaced by a copy of some finite graph.

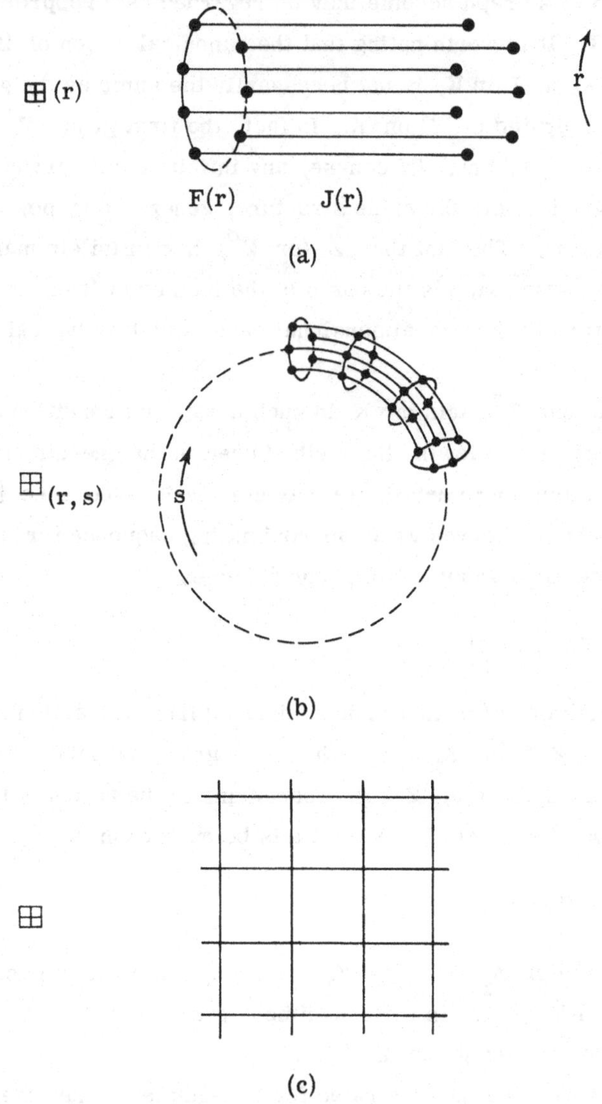

Fig. 20. The two-dimensional construction

5.2 The eigenvalue method in two dimensions

Let us consider what happens when the eigenvalue method, outlined in Chapter 2, is applied to a two-dimensional graph K. We have shown how to set up an approximating sequence of finite graphs $K_{(r, s)}$, where

$K_{(r,s)}$ is the sth cyclic graph associated with a graph scheme $K(r)$. If $T(r)$ is the transfer matrix of the scheme $K(r)$ with respect to some model $\mathfrak{M}$, then Theorem 2 tells us that

$$Z(\mathfrak{M}, K_{(r,s)}) = \text{trace}[T(r)]^s ,$$

which is expressible in terms of the eigenvalues of $T(r)$. Furthermore, the correlation functions and their limits may be investigated along the lines of Section 2.4, in the following way.

Let ω be a state on $K_{(r,s)}$, and let x, y be any two vertices of the fundamental graph $F(r)$ of $K(r)$. The correlation function $c_k^{(r)}(\omega)$ and its expected value $< c_k^{(r)} >_s$ in $K_{(r,s)}$ are defined just as in Section 2.4. We also have the quantities

$$\gamma_k(r) = \lim_{s\to\infty} < c_k^{(r)} >_s ,$$

$$\gamma_\infty(r) = \lim_{k\to\infty} \gamma_k(r) .$$

For any positive model, the argument presented in Theorem 4 explains how the fact that the maximum eigenvalue $\lambda_1(r)$ of $T(r)$ has unit multiplicity leads to the conclusion that $\gamma_\infty(r) = 0$. However, we are concerned here with a double sequence $K_{(r,s)}$, and so it is necessary to take the limit as $r \to \infty$ before letting $k \to \infty$. It is possible that, even though the next largest eigenvalue $\lambda_2(r)$ is strictly less than $\lambda_1(r)$ for every value of r, the ratio $\lambda_1(r)/\lambda_2(r)$ may approach unity as $r \to \infty$. If this 'asymptotic degeneracy' occurs, then the order parameter

$$\gamma_\infty(K) = \lim_{k\to\infty} \lim_{r\to\infty} \lim_{s\to\infty} < c_k^{(r)} >_s$$

may be non-zero, even for positive models.

In fact, the famous results of Onsager (see references for Chapter 1) show that this phenomenon does occur in the case of the infinite square lattice ⊞ . The schemes ⊞(r) and the graphs $⊞_{(r,s)}$ are as depicted in Fig. 20. The relevant models are the Ising models $\mathcal{I}[T]$, where T is the temperature. It is convenient to denote the 'conjugate' temperature, as defined in (3.5.5), by T^*; writing $\nu = L/kT$ and $\nu^* = L/kT^*$ an alternative form of (3.5.5) is

$$\sinh 2\nu \ \sinh 2\nu^* = 1.$$

Onsager showed that the two largest eigenvalues of the relevant transfer matrix $T(r)$ are as follows:

$$\lambda_1(r) = \exp \tfrac{1}{2}(\theta_1 + \theta_3 + \ldots + \theta_{2r-1})$$
$$\lambda_2(r) = \exp \tfrac{1}{2}(\theta_0 + \theta_2 + \ldots + \theta_{2r-2}),$$

where $\theta_0 = 2(\nu - \nu^*)$ and θ_l $(1 \leq l \leq 2r-1)$ is given by

$$\cosh \theta_l = \cosh 2\nu \cosh 2\nu^* - \cos(l\pi/r).$$

For $l \geq 1$ and large r, $\theta_{2l} \approx \theta_{2l+1}$, and so the limiting behaviour of the ratio $\lambda_1(r)/\lambda_2(r)$ depends on the limit of $\exp \frac{1}{2}(\theta_0 - \theta_1)$. Now θ_0 changes sign at $\nu = \nu^*$, while θ_1 does not. Consequently, the ratio approaches 1 if $\nu > \nu^*$ and we have asymptotic degeneracy of the eigenvalues. However, if $\nu < \nu^*$ the maximum eigenvalue remains isolated as $r \to \infty$.

In practical terms, this means that for low temperatures ($\nu > \nu^*$ implies $T < T_c$) there may be long-range order, whereas for sufficiently high temperatures ($T > T_c$) the order parameter is zero. Since such behaviour is similar to that exhibited by real magnetic materials, Onsager's result is a remarkable vindication of the Ising model of magnetism.

5.3 Existence of long-range order

Instead of seeking an explicit expression for the order parameter, we may attempt to prove the existence of long-range order by indirect means. That is the approach to be adopted in this section. We shall obtain a sufficient condition, involving the structure of a graph G, for the existence of long-range order on G with respect to a simple kind of positive model.

We shall discuss the case when each particle has just two possible configurations. For any state ω in $\Omega(G, A_2)$, and any pair (u, v) of vertices of G, we define the correlation function

$$c_{(u,\ v)}(\omega) = \begin{cases} 1 \text{ if } \omega(u) = \omega(v); \\ -1 \text{ if } \omega(u) \neq \omega(v). \end{cases} \qquad (5.3.1)$$

This is slightly more general than the correlation functions of Section 2.4. If $\mathcal{P} = (A_2, i)$ is a positive model, with weight function I, then for each pair (u, v) we have the expected value of $c_{(u,v)}$,

$$< c_{(u,v)} > = (1/Z) \sum_{\omega\in\Omega} c_{(u,v)}(\omega) I(\omega) .$$

We introduce a new measure of the 'order' in G, with respect to $\mathcal{P}$, defined by

$$\bar{\gamma}(G) = \frac{1}{|VG|^2} \sum_{(u,v)} < c_{(u,v)} > . \qquad (5.3.2)$$

If $\{G_n\}$ is a sequence of finite graphs we are interested in the limiting behaviour of $\bar{\gamma}(G_n)$ as $n \to \infty$. In order to avoid questions about the existence of the limit, we shall express our arguments in terms of bounds.

We begin with an alternative form for $\bar{\gamma}(G)$. Suppose that $s_v(\omega)$ is +1 or -1 according as $\omega(v) = 1$ or 0. Then

$$\sum_{(u,v)} c_{(u,v)}(\omega) = \sum_{(u,v)} s_u(\omega) s_v(\omega)$$

$$= \{\sum_u s_u(\omega)\}^2 = \{n_1(\omega) - n_0(\omega)\}^2 = M(\omega)^2,$$

where $n_1(\omega)$ and $n_0(\omega)$ denote the number of vertices x with $\omega(x) = 1$, 0 respectively, and $M(\omega) = n_1(\omega) - n_0(\omega)$ is a measure of the total 'magnetisation'. Taking expected values and using (5.3.2) we obtain

$$|VG|^2 \bar{\gamma}(G) = < M^2 > .$$

Now the Cauchy-Schwarz inequality implies that $< M^2 > \geq < |M| >^2$. Consequently, if we are given a sequence $\{G_n\}$, and we can find a constant C $(0 < C < 1)$ such that

$$< |M| > \geq C |VG_n| \text{ for all } n, \qquad (5.3.3)$$

then it follows that $\bar{\gamma}(G_n) \geq C^2$ for all n. This is the basis of our approach.

The possibility of satisfying (5.3.3), in the case of a positive model $\mathcal{P} = (A_2, i)$, depends on the ratio $\rho = i_0/i_1$. When $\rho = 1$ all states are equally probable and $<|M|>$ is identically zero. However, if $\rho > 1$ then there is bias towards like configurations at adjacent vertices. We have to show that, when the bias is sufficiently large, its effects will persist at great distances - or rather, that there are some sequences of graphs for which this is so.

We shall make a couple of assumptions to simplify the calculations. First, we shall only consider graphs with an odd number of vertices. This means that the set $\Omega = \Omega(G, A_2)$ is partitioned into sets Ω_0 and Ω_1, where Ω_0 contains those states ω for which $n_0(\omega) > n_1(\omega)$, and Ω_1 contains those for which $n_1(\omega) > n_0(\omega)$. In this case we have

$$\begin{aligned}<|M|> &= (1/Z) \sum_{\Omega} |n_1(\omega) - n_0(\omega)| \, I(\omega) \\ &= (1/Z) \{ \sum_{\Omega_0} [|VG| - 2n_1(\omega)] I(\omega) + \sum_{\Omega_1} [|VG| - 2n_0(\omega)] I(\omega) \} \\ &= |VG| - [<n_1>_0 + <n_0>_1], \qquad (5.3.4)\end{aligned}$$

where $<n_1>_0$ denotes the expected value of n_1, given that $n_0 > n_1$. (If $|VG|$ is even, the formula (5.3.4) is slightly more complicated.)

In order to estimate $<n_1>_0$, we introduce the notion of a 'domain'. Let D be a subset of VG such that the vertex-subgraph $\langle D \rangle$ is connected. We say that D is a **domain** (with respect to a state ω) if $\omega(v) = 1$ for each v in D, and $\omega(u) = 0$ for all vertices adjacent to D but not in D. Define

$$X_D(\omega) = \begin{cases} 1 & \text{if D is a domain w.r.t. } \omega; \\ 0 & \text{if not.} \end{cases}$$

Then, for $\omega \in \Omega_0$,

$$n_1(\omega) = \sum |D| \, X_D(\omega),$$

where the sum is over all connected subsets D of VG with $|D| < \frac{1}{2}|VG|$

At this point it is convenient to make a second simplifying assumption: we suppose that G is vertex-transitive, so that all its vertices are alike. This enables us to express $<n_1>_0$ as follows:

$$
\begin{aligned}
<n_1>_0 &= \sum_D |D| <X_D>_0 \\
&= \sum_D \sum_{x \in D} <X_D>_0 \\
&= \sum_x \sum_{D \ni x} <X_D>_0 \\
&= |VG| \sum_{D \ni z} <X_D>_0 ,
\end{aligned}
$$

where z is any specified vertex of G.

Substituting this expression for $<n_1>_0$ in (5. 3. 4), we see that the condition (5. 3. 3) will be satisfied if there is a constant A $(0 < A < \frac{1}{2})$ such that

$$\sum <X_D>_0 \leq A , \qquad (5.3.5)$$

and the constant A is the same for all graphs in the sequence $\{G_n\}$. The sum is now to be taken over all those connected subsets D of VG_n which contain some given vertex and satisfy $|D| < \frac{1}{2}|VG_n|$.

For a given subset D, we shall estimate $<X_D>_0$ as follows. Suppose that D is a domain with respect to a state ω in Ω_0. We may define a new state ω_D by changing the value of ω at all vertices in D from 1 to 0: since ω is in Ω_0, so is ω_D. If b(D) denotes the number of edges which border D, that is, which join vertices in D to vertices not in D, then the weights of ω and ω_D are related by

$$I(\omega)/I(\omega_D) = (i_1/i_0)^{b(D)} = \rho^{-b(D)} .$$

Hence, denoting by Σ' a sum over those states ω in Ω_0 for which D is a domain, we have

$$
\begin{aligned}
<X_D>_0 &= \Sigma' I(\omega) / \sum_{\Omega_0} I(\omega) \\
&\leq \Sigma' I(\omega) / \Sigma' I(\omega_D) \\
&= \rho^{-b(D)} . \qquad (5.3.6)
\end{aligned}
$$

Let $d_b(G)$ denote the number of connected subsets D of VG which contain a given vertex of G and satisfy $b(D) = b$, $|D| < \frac{1}{2}|VG|$. Using (5.3.6) we see that our requirement (5.3.5) becomes

$$\sum d_b(G_n)\rho^{-b} \leq A \text{ for all } n. \qquad (5.3.7)$$

If the coefficients $d_b(G_n)$ behave suitably, then it is possible that the sum can be bounded by making ρ sufficiently large. But first, we shall give an example of a sequence $\{G_n\}$ for which the method does not work. We recall from Chapter 2 that one-dimensional graphs (with respect to positive models) exhibit no long-range order. Let us therefore consider the circuit graphs, taking $G_n = C_{2n+1}$ to comply with the simplifying conditions imposed above. We need only consider the coefficient with $b = 2$ in order to see that (5.3.7) cannot hold. Any chain of r successive vertices of C_{2n+1} forms a connected set with $b = 2$, and it contributes to d_2 provided $r \leq n$. The number of such chains containing a given vertex is just r, so that

$$d_2(C_{2n+1}) = 1 + 2 + \ldots + n = \tfrac{1}{2}n(n-1).$$

Clearly, (5.3.7) cannot be satisfied for all n, however large ρ may be. The reason for our failure in this instance is that there is no upper bound for the size of a connected set which is bordered by just two edges, and this fact depends critically on the one-dimensional nature of the circuit graphs. Let us therefore pass on quickly to the two-dimensional lattice graphs.

Suppose $G_n = ⊞_{2n+1,\,2n+1}$, in the notation of the previous sections. If we confine our attention to a neighbourhood of some specified vertex z, then G_n may be considered to be planar. Given a connected set D containing z, the edges which border D correspond, in the dual graph, to a circuit of edges surrounding z. The number of such circuits with b edges is the number of connected sets D contributing to d_b. Now the first edge may be chosen from any of the b^2 edges within distance $\frac{1}{2}b$ of z, and each subsequent edge may be chosen in three ways. Every possible circuit is thereby counted b times, and many non-circuits are also counted, so that

$$d_b \le b.\, 3^{b-1}.$$

Comparing this with the left-hand-side of (5. 3. 7), we see that the infinite series $\Sigma d_b \rho^{-b}$ is convergent for $\rho > 3$, and by choosing ρ sufficiently large the sum can be made less than $\frac{1}{2}$.

The preceding argument is not quite complete, since we have ignored the possibility of connected sets which, instead of being planar, are 'cylindrical' in form. Fortunately, the number of edges required to border such a set is at least $4n + 2$, from which it follows that the number of these sets contributing to $d_b(G_n)$, being bounded by a function of n, is also bounded by a function of b. Hence the argument given is essentially correct. In summary, we have shown that, for this particular sequence $\{G_n\}$, the quantities $\bar{\gamma}(G_n)$ are bounded away from zero when the ratio ρ is large enough.

For the Ising model, $\rho = \exp(2L/kT)$, and so we have established the existence of long-range order (permanent magnetism) at low temperatures, for the square lattice. Similar arguments may be used for other graphs with dimensionality $d > 2$, but it would be preferable to set up a more direct link between the definition of dimensionality and the criterion (5. 3. 7).

It would be idle to pretend that the material treated in this book is in its final form; clearly, there are many difficulties to be overcome before the correct conceptual framework can be constructed. Some of the ideas presented here are inelegant, inconsistent, and, quite possibly, wrong. If some of my errors stimulate other mathematicians to put things right, then the book will have served its purpose.

NOTES AND REFERENCES FOR CHAPTER 5

There is some evidence that certain parameters associated with phase transitions - for example, the so-called critical exponents - depend only on the model and the dimensionality of the graph, not on the graph itself. This is the substance of the 'Universality Hypothesis' stated by Kadanoff; see [4]. Apparently, there is a close relationship between such notions and the technique of 'renormalization' [1], but no complete mathematical theory has yet appeared. The very notion of dimensionality is sometimes used rather vaguely, and confused with other ideas, such as planarity. Our definition makes it clear that a two-dimensional graph need not be planar.

A general treatment of the existence of the thermodynamic limit, based on our definition of dimensionality, is given by Grimmett [3].

Onsager's results for the Ising model on the square lattice have subsequently been derived in many different ways. Temperley [5] surveys the various techniques, and remarks on the disappointing fact that little new information has emerged from them.

The basis of the argument in Section 5.3 is due to Peierls (reference [6] of Chapter 1). His work was concerned with the Ising model, and it was the first indication that the model might be more useful than Ising's original failure in one dimension had suggested. Griffiths [2] corrected some details in the Peierls argument. The aim of our treatment is to clear the way for a formulation based on the idea of dimensionality.

1. Barber, M. N. An introduction to the fundamentals of the renormalization group in critical phenomena. Physics Reports, 29, no. 1 (1977), 1-84.
2. Griffiths, R. B. Peierls proof of spontaneous magnetization in a two-dimensional Ising ferromagnet. Phys. Rev. 136A (1964), 437-9.
3. Grimmett, G. R. (In preparation.)
4. Kadanoff, L. Scaling, universality and operator algebras. Phase Transitions and Critical Phenomena (C. Domb and M. S. Green, eds.), Volume 5, pp. 1- .
5. Temperley, H. N. V. Two-dimensional Ising models. Phase Transitions and Critical Phenomena (C. Domb and M. S. Green, eds.), Volume 1, pp. 227-67.

Appendix A: Distributive identities

The inclusion of this appendix is justified on grounds of utility, rather than depth.

Lemma I. *Let A and B be finite sets and θ a complex-valued function defined on $A \times B$. Let B^A denote the set of functions $f : A \to B$. Then*

$$\sum_{f \in B^A} \prod_{a \in A} \theta(a, f(a)) = \prod_{a \in A} \sum_{b \in B} \theta(a, b) .$$

Proof. For each a in A let

$$S(a) = \sum_{b \in B} \theta(a, b) .$$

The product of the sums $S(a)$ may be simplified by using the distributive law ('multiplying out') which yields a sum of product terms. Each product term has one factor from each sum, that is, one factor $\theta(a, b)$ for each a in A. The choice of factors corresponds to defining a function f which assigns one element $b = f(a)$ to each a in A. Hence the result. //

Lemma II. *Let A be a finite set, N the set of non-negative integers, and θ a complex-valued function defined on $A \times N$. Suppose that, for each a in A, the series $\Sigma\theta(a, n)$ is absolutely convergent. Let N^A denote the set of functions $f : A \to N$ and let E be an enumeration of N^A (that is, $E : N \to N^A$ is a one-to-one correspondence).*

For each f in N^A define

$$\pi(f) = \prod_{a \in A} \theta(a, f(a)) ,$$

and let $\pi_n = \pi(E(n))$. Then the series $\Sigma\pi_n$ is absolutely convergent and

$$\sum_{n=0}^{\infty} \pi_n = \prod_{a \epsilon A} \sum_{n=0}^{\infty} \theta(a, n) .$$

Proof. When $|A| = 2$ the result is equivalent to theorems on double series, proved in elementary analysis courses. A presentation closely akin to our formulation may be found on pages 372-5 of the text by Apostol (reference below). For $|A| > 2$, one may either modify the standard proofs, using Lemma I where needed, or proceed by induction on $|A|$. //

Note. Since the result is true for any enumeration of the set N^A, it is permissible to write the left-hand-side as a sum over functions f from A to N. This is the form required in Section 4.4, where A is the finite set St_N.

REFERENCE

Apostol, T. M. Mathematical analysis. (Addison-Wesley, Reading (Mass.), 1957.)

Appendix B: The Perron-Frobenius theorem

We shall prove the theorem in its simplest form. Generalizations may be found in the references at the end of this appendix.

Theorem. Let $A = (a_{ij})$ be an $n \times n$ matrix whose entries are real and positive. Then A has a real, positive eigenvalue μ with the following properties:

(a) there is a positive eigenvector z associated with μ;

(b) if λ is any other eigenvalue of A, then $|\lambda| < \mu$;

(c) any eigenvector associated with μ is a multiple of z.

Proof. Let S denote the compact subset of $\mathbf{R}^n$ consisting of those vectors x which satisfy $x > 0$ and $x'x = 1$. Define a function $\phi : S \to \mathbf{R}$ by the rule

$$\phi(x) = \min (Ax)_j / x_j,$$

where the minimum is taken over those j for which $x_j \neq 0$. Since $x'x = 1$, not all x_j are zero and ϕ is well-defined. Furthermore, it is upper semi-continuous and bounded above (by the maximum row sum of A). Hence ϕ attains its supremum μ on S, and there is some z in S for which

$$\phi(z) = \sup_{x \in S} \phi(x) = \mu .$$

By the defining property of the supremum,

$$(Az)_j \geq \mu z_j \qquad (j = 1, 2, \ldots, n), \qquad (*)$$

that is, $Az \geq \mu z$. Writing $y = Az$, we have $y - \mu z \geq 0$.

If $y - \mu z \neq 0$, then $A(y - \mu z) > 0$, since every entry of A is positive. But this gives $Ay > \mu y$, and consequently $\phi(\hat{y}) > \mu$, where $\hat{y}$

is obtained by normalizing y (so that $\hat{y}$ belongs to S). This contradicts the definition of μ. Hence $y - \mu z = 0$, that is, $Az = \mu z$. We have shown that μ (which, by its definition, is real and positive) is an eigenvalue of A with associated eigenvector z. Also, since z is in S, we have $z \geq 0$, $z \neq 0$, which gives $Az > 0$ and $\mu z > 0$; that is, $z > 0$. Statement (a) is therefore proved.

Suppose λ is any eigenvalue of A, and $Ax = \lambda x$ for some $x \neq 0$. Both λ and x may be complex. Since $(Ax)_j = \lambda x_j$ and $|a_{ij}| = a_{ij}$,

$$|\lambda|\ |x_j| = |\Sigma a_{ij} x_i| \leq \Sigma a_{ij} |x_i| .$$

So if x_+ denotes the vector obtained from x by taking the modulus of each component, we have

$$|\lambda| \leq (Ax_+)_j/(x_+)_j \quad \text{for each } j,$$

and normalizing to obtain a vector in S with the same property, we deduce that $|\lambda| \leq \mu$.

We shall show that if $|\lambda| = \mu$, then $\lambda = \mu$. Now, when $|\lambda| = \mu$,

$$(Ax_+)_j \geq |\lambda|(x_+)_j = \mu(x_+)_j ,$$

which is just the situation in (*) above, with x_+ instead of z. By the same arguments as we used there, it follows that $Ax_+ = \mu x_+$, that is, $Ax_+ = |\lambda| x_+$. But we began with $Ax = \lambda x$, so

$$|(Ax)_j| = |\lambda|\,|x_j| = |\lambda|(x_+)_j = (Ax_+)_j ,$$

that is,

$$|\Sigma a_{ij} x_i| = \Sigma a_{ij} |x_i| .$$

Consequently, since the coefficients a_{ij} are real, it follows that all the complex numbers x_i must have equal arguments, and $x_i = K|x_i|$ for some complex constant K. Thus x is a multiple of x_+, and (since $Ax_+ = \mu x_+$) $\lambda x = Ax = \mu x$, whence $\lambda = \mu$. Statement (b) is proved.

Finally, suppose that w is any vector satisfying $Aw = \mu w$. Then for any complex number γ, the vector $v = w - \gamma z$ is also an eigenvector associated with μ, provided that $v \neq 0$. Let v_+ be obtained from v by the construction given above. Arguing as before, we find that if $v_+ \neq 0$, then $v_+ > 0$. But this cannot be true for all γ, since we can surely choose γ to make one component of v_+ zero. Thus for some γ, $v = 0$; that is, w is a multiple of z. This completes the proof. //

REFERENCES

Gantmacher, F. R. Applications of the theory of matrices, Vol. II. (Chelsea, New York, 1959.)

Seneta, E. Non-negative matrices. (George Allen and Unwin, London, 1973.)

Index